Library of
Davidson College

# DEGREES OF UNSOLVABILITY

Dedicated to

S.C. KLEENE,

who made recursive function theory into a theory.

# NORTH-HOLLAND
# MATHEMATICS STUDIES    2

# Degrees of Unsolvability

**JOSEPH R. SHOENFIELD**
*Professor of Mathematics*
*Duke University*
*Durham, N.C., USA*

1971

NORTH-HOLLAND PUBLISHING COMPANY AMSTERDAM-LONDON
AMERICAN ELSEVIER PUBLISHING COMPANY, INC. - NEW YORK

© NORTH-HOLLAND PUBLISHING COMPANY – AMSTERDAM – 1971

All Rights Reserved. No part of this publication may be reproduced, stored in a retrieval system, or transmitted, in any form or by any means, electronic, mechanical, photocopying, recording or otherwise, without the prior permission of the Copyright owner.

ISBN North-Holland 07204 2061 x
ISBN American Elsevier 00444 10128 4

PUBLISHERS:

NORTH-HOLLAND PUBLISHING COMPANY – AMSTERDAM
NORTH-HOLLAND PUBLISHING COMPANY, LTD. – LONDON

SOLE DISTRIBUTORS FOR THE U.S.A. AND CANADA:

AMERICAN ELSEVIER PUBLISHING COMPANY, INC.
52 VANDERBILT AVENUE
NEW YORK, N.Y. 10017

PRINTED IN THE NETHERLANDS

# CONTENTS

## DEGREES OF UNSOLVABILITY - J.R. Shoenfield

| | Page |
|---|---|
| Introduction | VII |
| 0 - Terminology and Notation | 1 |
| 1 - Recursive Functions | 3 |
| 2 - Isomorphisms | 8 |
| 3 - Algorithms | 12 |
| 4 - Relative Recursiveness | 15 |
| 5 - Recursive Enumerability | 20 |
| 6 - Degrees | 26 |
| 7 - Evaluating Degrees | 31 |
| 8 - Incomparable Degrees | 39 |
| 9 - Upper and Lower Bounds | 43 |
| 10 - The Jump Operation | 46 |
| 11 - Minimal Degrees | 49 |
| 12 - Simple Sets | 57 |
| 13 - The Priority Method | 60 |
| 14 - The Splitting Theorem | 66 |
| 15 - Maximal Sets | 72 |
| 16 - Infinite Injury | 83 |
| 17 - Index Sets | 93 |
| 18 - Branching Degrees | 99 |

## INTRODUCTION

These notes originated from a seminar which I gave at UCLA in 1967. Several graduate students in the seminar wrote up a set of notes. After lecturing from these notes at the Catholic University of Santiago in 1969, I rewrote the notes in the present form.

A set of notes should have some purpose other then collecting theorems from the literature. My main purpose here is to show that proofs in degree theory need not be as complicated and incomprehensible as they appear to be in the literature. (Some of the later articles of Friedberg and Lachlan provide pleasant exceptions.) The main difference between the proofs here and those in the literature is that I have directly given the idea behind the proof, instead of translating it into an obscure language from which the reader must translate back. I have also avoided introducing too much notation and carefully avoided introducing specific enumerations of pairs, finite sets, finite sequences, etc. I am convinced that if the experts in degree theory would follow these precepts in writing their articles, other logicians would read these articles instead of merely looking at them with awe (or disgust). Mea culpa.

I have also tried to show how little about recursive functions is needed. I have not even introduced what is usually called a 'rigorous' definition of a recursive function. Of course, such a definition is needed for some applications of recursive function theory, but not for the theory of degrees. Besides simplifying the exposition, I hope this may indicate what sort of axioms are needed for an axiomatic theory of recursive functions.

No attempt has been made to cover even the topics treated in a complete manner. The main object has been to give examples of the techniques which have proved most useful.

This book owes much to discussions with several people, particularly Alastair Lachlan, Hartley Rogers, Gerald Sacks, and Mike Yates. The NSF has provided valuable financial support.

J. R. Shoenfield

Durham, N. C.
August 10, 1971

## 0. Terminology and Notation

We use the following logical notation: $\neg$ for _not_; $\vee$ for _or_; $\&$ for _and_; $\rightarrow$ for _implies_; $\leftrightarrow$ for _iff_ (= _if and only if_). We call $\neg$, $\vee$, $\&$, $\rightarrow$ and $\leftrightarrow$ _connectives_.

If $P(x)$ is a statement about $x$, $\exists x\, P(x)$ means that $P(x)$ holds for some $x$ and $\forall x\, P(x)$ means that $P(x)$ holds for all $x$. This notation will only be used when $x$ is restricted to vary through some fixed set. Sometimes this set is directly indicated by the notation. Thus $(\exists x \in y)\, P(x)$ means that $P(x)$ holds for some $x$ in $y$; and $(\forall x < y)\, P(x)$ means that $P(x)$ holds for every $x$ such that $x < y$. We call $\exists x$ and $\forall x$ _quantifiers_; the former is an _existential_ quantifier and the latter is a _universal_ quantifier.

We use _mapping_ and _class_ with their usual meaning; _function_ and _set_ will be given special meanings. We use $\emptyset$ for the empty class. We do _not_ identify a mapping with a class of ordered pairs. If F is a mapping of A onto B, then A is the _domain_ of F and B is the _range_ of F. If F is one-one, the inverse of F is designated by $F^{-1}$.

We use the notation $\langle x_1, \ldots, x_n \rangle$ for ordered n-tuples. An ordered n-tuple is _not_ identified with any class.

A _natural number_ is a non-negative integer. We use N for the class of natural numbers, and use i, j, k, m, n, r, s, and t to designate natural numbers.

An infinite sequence is identified with a mapping having N as its domain. As usual, we write $x_n$ for the value of the infinite sequence x at the argument n, and designate the infinite sequence by $\{x_n\}$.

A finite sequence of length n is identified with a mapping with domain $\{0, 1, \ldots, n-1\}$ (but not with an ordered n-tuple). Again $x_i$ is the value of the sequence x at the argument i. The length of a finite sequence x is designated by lh(x). We use $\emptyset$ for the sequence of length 0 (as well as for the empty class).

## 1. Recursive Functions

Recursion theory is the abstract theory of computations. All of the computations which we consider will be, at least in theory, performable in a finite length of time. This means that the objects with which we compute must be __finite objects__, i.e., objects which can be specified by a finite amount of information.

Some examples will clarify the notion of a finite object. A natural number is a finite object, since it can be specified by giving the Arabic numeral designating that number. On the other hand, a real number is generally not a finite object, since to specify it we must, say, give each of its infinitely many decimal places. An n-tuple of finite objects is a finite object, since it can be specified by specifying each of the n objects in turn. A finite class of finite objects is a finite object; an infinite class of finite objects is generally not a finite object. A finite sequence of finite objects is a finite object; an infinite sequence generally is not.

A __space__ is an infinite class X of finite objects such that, given a finite object x, we can decide whether or not x belongs to X. We give some examples of spaces.

(1) The class N of natural numbers is a space.

(2) If X and Y are spaces, then $X \times Y$ is a space.

(3) If X is a space, then the class $Sub(X)$ of finite subclasses of X is a space.

(4) If X is either a space or a finite non-empty class of finite objects, then the class Sq(X) of finite sequences of elements of X is a space.

We use X, Y, and Z for spaces. Generally x, y, and z are elements of X, Y, and Z respectively.

A <u>function from</u> X <u>to</u> Y is a mapping from X to Y; a <u>set in</u> X is a subclass of X. A <u>function</u> is always a function from a space to a space, and a <u>set</u> is always a set in a space. We use F, G, H, L and M for functions and A, B, C, D and E for sets.

A function F from a Cartesian product of n spaces is called a <u>function of</u> n <u>arguments</u>; we write simply $F(x_1, \ldots, x_n)$ for $F(\langle x_1, \ldots, x_n \rangle)$. A set in a Cartesian product of n spaces is called a <u>relation of</u> n <u>arguments</u>. If we refer to ..x..y.. <u>as a function of</u> x,y, we mean the function F defined by $F(x,y) = $ ..x..y.. . If we refer to ..x..y.. <u>as a relation of</u> x,y, we mean the relation A defined by $\langle x,y \rangle \in A \leftrightarrow $ ..x..y.. . By the relation =, we mean x = y as a relation of x,y; similarly the relation ⊂, the relation <, etc.

It is convenient to identify each set with a function, so that the definitions we give for functions will apply also to sets. We identify the set A in X with the function from X to N which takes the value 1 for arguments in A and takes the value 0 for arguments not in A. As a mnemonic, we write Tr for 1 and Fa for 0. Thus $A(x) = $ Tr iff $x \in A$ is true, and $A(x) = $ Fa iff x ∈ A is false.

A function F from X to Y is _recursive_ if there is an algorithm by which, given a member x of X, we can compute $F(x)$. It follows that a set A in X is recursive iff there is an algorithm by which, given a member x of X, we can compute whether x ∈ A or x ∉ A.

We give some examples of recursive functions and sets. The addition and multiplication functions are recursive functions from N×N to N; algorithms for these functions are taught in elementary school arithmetic. If $F(0) = 2$ and $F(n)$ for $n \neq 0$ is the nth decimal place in the real number e, then F is a recursive function from N to N; here the algorithm comes from the theory of infinite series. The set of primes is a recursive set in N; the algorithm is the sieve of Eratosthanes.

The identity function from X to X and the projections of X×Y on X and Y are recursive. The length function lh is a recursive function from Sq(X) to N. The relation = in X×X and the relation ⊂ in Sub(X)×Sub(X) are recursive. The relations <, >, ≤, and ≥ in N×N are recursive.

Now we describe some methods for obtaining recursive functions and sets from other recursive functions and sets. The composition of two recursive functions is recursive; for we can compute $(F \circ G)(x)$ by computing $G(x) = y$ and then computing $F(y)$. The same holds for any composition of functions of several arguments. Thus if F is defined by

$$F(x,y) = G(H(y,x), L(x), y)$$

where G, H, and L are recursive, then F is recursive.

If A is defined by $x \in A \leftrightarrow F(x) \in B$ where F and B are recursive, then A is recursive. Again this extends to several arguments. Thus if A is defined by

$$\langle x,y \rangle \in A \leftrightarrow \langle F(x), G(y,x), y \rangle \in B$$

where F, G, and B are recursive, then A is recursive.

The union, intersection, or set difference of two recursive sets (in the same space) is recursive. The <u>complement</u> of a set A in X, designated by $A^c$, is the set $X - A$; it is recursive iff A is recursive.

Any combination of recursive sets by means of connectives is recursive. Thus if A is defined by

$$x \in A \leftrightarrow (x \in B \rightarrow x \in C) \,\&\, x \in D$$

where B, C, and D is recursive, then A is recursive. The same is not true when we use quantifiers. Thus suppose that we define

$$x \in A \leftrightarrow \exists y (\langle x,y \rangle \in B)$$

where B is recursive. To test whether $x \in A$ or $x \notin A$, we must test whether $\langle x,y \rangle \in B$ or $\langle x,y \rangle \notin B$ for each y. Since there are infinitely many y, this cannot be done in a finite length of time.

This problem does not arise if the quantifier varies through a finite set. Thus if A is defined by

$$\langle x,k \rangle \in A \leftrightarrow (\exists n < k)(\langle x,n,k \rangle \in B)$$

where B is recursive, then A is recursive. Similar remarks apply to the quantifiers $(\forall n < k)$, $(\exists n \leq k)$, and $(\forall n \leq k)$; and to the quantifiers $(\exists x \in y)$ and $(\forall x \in y)$ where $Y = \mathrm{Sub}(X)$.

We let $\mu n(..n..)$ be the smallest n such that ..n..; if there is no n such that ..n.., then $\mu n(..n..)$ is undefined. If A is recursive and $F(x) = \mu n(\langle x,n \rangle \in A)$ for all x, then F is recursive; for we may compute $F(x)$ by examining $\langle x,0 \rangle$, $\langle x,1 \rangle$, ... in turn until we come to the first one which belongs to A.

## 2. Isomorphisms

An <u>isomorphism</u> of X and Y is a one-one function F from X onto Y such that F and $F^{-1}$ are recursive. If such an isomorphism exists, we say that X and Y are <u>isomorphic</u>.

We shall consider only properties of spaces which are invariant under isomorphisms. For example, we show that recursiveness of function is invariant under isomorphisms. Let F be a recursive function from X to Y. Let G be an isomorphism of X and X', and let H be an isomorphism of Y and Y'. The function F' from X' to Y' which corresponds to F is given by the commutative diagram

$$\begin{array}{ccc} X & \xrightarrow{F} & Y \\ G \downarrow & & \downarrow H \\ X' & \xrightarrow{F'} & Y' \end{array}$$

Thus $F' = H \circ F \circ G^{-1}$. Since H, F, and $G^{-1}$ are recursive, F' is recursive.

<u>Isomorphism Theorem</u>. Any two spaces are isomorphic.

As a first step towards the proof, we note that the identity mapping from X to X is an isomorphism; that the inverse of an isomorphism is an isomorphism; and that the composition of two isomorphisms is an isomorphism. It follows that the relation of being isomorphic is an equivalence relation. Hence we need only show that for every X, N is isomorphic to X.

A <u>listing</u> of a set A in X is a one-one recursive function from N to X with range A. Otherwise stated, a listing of A is an infinite sequence $\{x_n\}$ of elements of A in which every member of A appears exactly once such that given n, we can compute $x_n$.

Lemma 1. If F is a listing of X, then F is an isomorphism of N and X.

Proof. We need only show that $F^{-1}$ is recursive. Given x, we compute $F(0)$, $F(1)$, ... until we find an n such that $F(n) = x$. Then $F^{-1}(x) = n$. Q.E.D.

Lemma 2. If A is an infinite set in X, and F is a recursive function from N to X with range A, then A has a listing.

Proof. Define a function G from N to X inductively as follows: $G(n) = F(m)$, where m is the smallest number such that $F(m)$ is distinct from each of $G(0)$, $G(1)$, ..., $G(n-1)$. This m exists because A is infinite. Clearly G is a one-one function from N to X with range A. Since F is recursive, we can compute $G(n)$ when $G(0)$, $G(1)$, ..., $G(n-1)$ are known. Hence G is recursive. Q.E.D.

Lemma 3. If X has a listing, then any space Y included in X has a listing.

Proof. Let F be a listing of X. Pick $y_0 \in Y$. Set $G(n) = F(n)$ if $F(n) \in Y$, and $G(n) = y_0$ otherwise. Then G is a recursive function from N to X with range Y; so Y has a listing by Lemma 2. Q.E.D.

If x is a finite object, we may write down a complete description of x. We may suppose that the symbols used in this description are chosen from a finite class $\Gamma$ independent of x.

(It would suffice to put into $\Gamma$ all symbols used in English, including punctuation marks, and all the usual mathematical symbols.) Since $\Gamma$ is a finite class of finite objects, $Sq(\Gamma)$ is a space; and all of our descriptions belong to this space.

Lemma 4. The space $Sq(\Gamma)$ has a listing.

Proof. Let $x_1$, $x_2$, ..., $x_r$ be the symbols in $\Gamma$. Let $F(0)$ and $F(1)$ be the empty sequence. If $n > 1$, let the prime power decomposition of n be $p_1^{n_1} p_2^{n_2} ... p_k^{n_k}$, where $p_1 < p_2 < ... < p_k$ and all of the exponents are positive. If all of the exponents are $\leq r$, let $F(n)$ be $x_{n_1} x_{n_2} ... x_{n_k}$; otherwise, let $F(n)$ be the empty sequence. Then F is a recursive function from N onto $Sq(N)$; so $Sq(N)$ has a listing by Lemma 2. Q.E.D.

Let X be a space. The definition of a space implies that given a member of $Sq(N)$, we can decide if it is a description of an object in X, and, if it is, we can find the object. Let Y be the set of descriptions of objects in X; and for $y \in Y$, let $F(y)$ be the object described by y. Then Y is a space included in $Sq(N)$ and F is a recursive function from Y onto X.

By Lemmas 4 and 3, Y has a listing G. Then F∘G is a recursive function from N onto X. By Lemma 2, X has a listing; so by Lemma 1, N is isomorphic to X. Q.E.D.

The Isomorphism Theorem states that all spaces are equivalent for our purposes. We often make tacit use of this fact. Thus we may state a result for all spaces and prove it only for N.

We suppose that a listing of $N \times N$ is fixed once and for all.  If the pair corresponding to n under this listing is $\langle i,j \rangle$, we say that n is in <u>row</u> i and <u>column</u> j.  (Thus we picture the listing as arranging the natural numbers in an infinite square array.)  Given n, we can compute its row and column; given i and j, we can compute the number in row i and column j.

## 3. Algorithms

The exact nature of an algorithm depends upon the method of computation. If the computation uses pencil and paper, the algorithm may be a set of directions in English; if the computation uses a computing machine, the algorithm may be a program for that machine. We shall examine some properties of algorithms which are independent of such considerations.

An algorithm for a function F from X to Y is a rule for making a computation leading from x to F(x). We call x the input and y the output of the computation. We think of the computation as taking place in steps. The algorithm tells us how to obtain the next step from the steps already completed and the input.

An algorithm from X to Y is a rule which may be applied to an input in X and a finite sequence of computation steps to obtain either a new computation step or an output in Y. The new step or output is obtained from the input and the given sequence of steps without making intermediate calculations, exercising any ingenuity, or supplying any further information.

An algorithm is a rule and hence a finite object; and the class Alg(X,Y) of algorithms from X to Y is a space. We use I, J, and K for algorithms.

Let I ∈ Alg(X,Y). Given an input x ∈ X, we may apply I to obtain a step, apply I to this to obtain another step, and so on. We then say that we are computing according to I. The process of obtaining new steps may go on forever, or it may terminate with an output in Y. The set of inputs for which we obtain an output is designated by $W_I$. If $x \in W_I$, the output obtained from the input x is designated by $[I](x)$, and the computation leading to it is called the computation of $[I](x)$. If $x \notin W_I$, $[I](x)$ is undefined.

We note that $[I]$ is a mapping from the set $W_I$ in X to Y. A mapping from a set in X to Y is called a partial function from X to Y. The letters reserved for functions will also be used to designate partial functions, but only when this is explicitly indicated. A partial function is total if it is a function. A partial function F from X to Y is recursive if there is an algorithm I from X to Y such that $F = [I]$. (If F is a function, this agrees with our previous definition.) Any such I is then called an algorithm for F.

We adopt a convention for equality between expressions (such as $[I](x)$) which may be undefined. We let $U = V$ hold if both U and V are defined and they have the same value, or if both U and V are undefined. In all other cases, $U \neq V$.

We can use the above to give an example of a non-recursive function. Let $\{I_n\}$ be a listing of Alg(N,N). Define a function F from N to N by

$$F(n) = [I_n](n) + 1 \quad \text{if } n \in W_{I_n},$$
$$= 0 \quad \text{otherwise.}$$

For each n, $F(n) \neq [I_n](n)$; so $F \neq [I_n]$ for all n. It follows that F is not recursive.

We can also show that $x \in W_I$ (where $I \in Alg(X,Y)$) is not a recursive relation of x,I. In view of the Isomorphism Theorem, it will suffice to prove this when $X = Y = N$. But then if $n \in W_I$ were a recursive relation of n,I, the F defined in the above paragraph would be recursive.

A similar proof shows that $[I](x) = y$ is not a recursive relation of I, x, y. Here the function F is defined by

$$F(n) = 1 \quad \text{if } [I_n](n) = 0,$$
$$\phantom{F(n)} = 0 \quad \text{otherwise.}$$

There are, however, some approximations to these relations which are recursive. Let $W_{I,n}$ be the set of x in $W_I$ such that the computation of $[I](x)$ has less than n steps. For $x \in W_{I,n}$, $[I]_n(x) = [I](x)$; for $x \notin W_{I,n}$, $[I]_n(x)$ is undefined.

We claim that $x \in W_{I,n}$ is a recursive relation of x, I, n and that $[I]_n(x) = y$ is a recursive relation of I, n, x, y. For let I, n, x, y be given. We compute according to I with the input x until either an output is obtained or n steps have been completed. Then $x \in W_{I,n}$ iff we obtain an output, and $[I]_n(x) = y$ iff we obtain the output y.

We note that $W_{I,n}$ and $[I]_n$ approximate $W_I$ and $[I]$ in the following sense. If $x \notin W_I$, then $x \notin W_{I,n}$ for all n. If $x \in W_I$, then there is an $n_0$ such that $x \notin W_{I,n}$ for $n < n_0$ and $x \in W_{I,n}$ and $[I]_n(x) = [I](x)$ for $n \geq n_0$.

## 4. Relative Recursiveness

In the computations considered so far, all of the information is supplied by the algorithm and the input. We now consider computations in which certain additional information may be used.

Let H be a function from N to N, and define F by $F(n) = H(2 \cdot H(n))$. If H is recursive, then F is recursive. Whether or not H is recursive, we can compute $F(n)$ from n if we are given an object which when it is given an argument k will supply the value $H(k)$. Such an object is called an <u>oracle</u> for H.

We extend the notion of an algorithm to allow the use of oracles. The extension consists of allowing a new instruction to appear in the algorithm. This new instruction tells us that the next computation step is to be the value given by the oracle for the argument obtained at the last computation step. This instruction would not have been allowed previously, since it requires us to obtain additional information.

To compute according to one of our extended algorithms, we need both an input and an oracle. We may then run into the following problem: we may be instructed to ask the oracle for the value corresponding to the argument z just obtained, and z may not be an object in the domain of the function for which we have an oracle. We make the convention that in this case, the next step is to be merely a repetition of the last previous step.

The same convention allows us to use the new type of algorithm to compute without an oracle. When the algorithm says to consult the oracle, we merely repeat the last previous step. With this convention, we make take all of our earlier references to algorithms to be to algorithms in the extended sense. This does not affect the validity of any of our results, and allows us to avoid dealing with two kinds of algorithms.

Let I be an algorithm from X to Y, and let H be a function. Then $W_I^H$ is the set of x such that we obtain an output when we compute according to I using the input x and an oracle for H. This output is then designated by $[I]^H(x)$, and the computation leading to it is called the computation <u>of</u> $[I]^H(x)$. If $x \notin W_I^H$, then $[I]^H(x)$ is undefined.

A partial function F from X to Y is recursive <u>in</u> H (or <u>relative to</u> H) if there is an algorithm I from X to Y such that $F = [I]^H$. Any such I is then called an algorithm <u>for</u> F <u>in</u> H.

Let $W_{I,n}^H$ be the set of $x \in W_I^H$ such that the computation of $[I]^H(x)$ has less than n steps. For $x \in W_{I,n}^H$, $[I]_n^H(x) = [I]^H(n)$; for $x \notin W_{I,n}^H$, $[I]_n^H(x)$ is undefined. Then $x \in W_{I,n}^H$, as a relation of x, I, n, is recursive in H, and $[I]_n^H(x) = y$, as a relation of I, n, x, y, is recursive in H. Moreover, $W_{I,n}^H$ and $[I]_n^H$ approximate $W_I^H$ and $[I]^H$ in the same sense as before.

What we have done so far is to replace <u>computation</u> by computation using an oracle in H in our previous results. This process is called relativization to H. The reason that it leads to correct results is that we have so far made no use of the fact that a computation uses no external source of information (and, in particular, no oracle). Since we will never make use of this fact, all of our results will remain valid upon relativization to H. We shall often make tacit use of this fact.

Every recursive function is recursive in H. For if F is recursive, then there is an algorithm I for F which makes no use of oracles; so $F = [I] = [I]^H$.

If F is recursive in H and H is recursive in G, then F is recursive in G. For suppose that we have an oracle for G. Given x, we start to compute $F(x)$ as if we had an oracle for H. However, when we are instructed to ask the oracle for a value $H(z)$, we instead compute $H(z)$, using our oracle for G. A similar proof shows that if F is recursive in H and H is recursive, then F is recursive.

Clearly H is recursive in H. More generally, if $F(x) = H(x)$ for all but a finite number of arguments x, then F is recursive in H. (The algorithm will list this finite number of arguments and the value of F at each of these arguments.)

Let H be a partial function from Z to some space. An oracle for H is an object which, given a z in Z, will give the value $H(z)$ if z is in the domain of H and will give no answer if z is not in the domain of H.

In computing according to an algorithm I with an oracle for a partial function, a new situation may arise: we may ask the oracle for a value and get no answer. In this case, we stop computing without obtaining an output. We may then define $W_I^H$, $[I]^H$, $W_{I,n}^H$, and $[I]_n^H$ for a partial function H as before.

Let H be a partial function and let $x \in W_I^H$. We say that z is <u>used</u> in the computation of $[I]^H(x)$ if during that computation the oracle is asked for the value $H(z)$. The following obvious but important fact is called the Use Principle: If H is a partial function such that $x \in W_I^H$, and G is a partial function such that $G(z) = H(z)$ for every z used in the computation of $[I]^H(x)$, then $x \in W_I^G$, $[I]^G(x) = [I]^H(x)$, and $[I]_n^G(x) = [I]_n^H(x)$ for all n.

If G and H are partial functions from X to Y, $G \subset H$ means that H is an extension of G. If $G \subset H$, then $[I]^G \subset [I]^H$ by the Use Principle.

A <u>finite function</u> from X to Y is a partial function from X to Y whose domain is finite. A finite function is a finite object, and the class of finite functions from X to Y is a space. We use $\sigma$ and $\pi$ for finite functions.

As a consequence of the Use Principle,

(1) $\qquad x \in W_I^H \leftrightarrow (\exists \sigma \subset H)(x \in W_I^\sigma)$

and

(2)  $[I]^H(x) = y \leftrightarrow (\exists \sigma \subset H)([I]^\sigma(x) = y).$

We also note that $x \in W^\sigma_{I,n}$ is a recursive relation of $x$, $\sigma$, $I$, $n$, and that $[I]^\sigma_n(x) = y$ is a recursive relation of $I$, $\sigma$, $n$, $x$, $y$. This is proved as before, noting that a knowledge of $\sigma$ enables us to determine what answers an oracle for $\sigma$ will give and when it will give no answer.

The elements of $Sq(\{Fa, Tr\})$ are called <u>strings</u>. We use $\alpha$, $\beta$, $\gamma$, and $\delta$ for strings. By the Use Principle, we have for $A$ a subset of $N$:

(3)  $x \in W^A_I \leftrightarrow (\exists \alpha \subset H)(x \in W^\alpha_I)$

and

(4)  $[I]^A(x) = y \leftrightarrow (\exists \alpha \subset H)([I]^\alpha(x) = y).$

We could extend the above to computations using several oracles; but it easier to replace several oracles by one. Let $H_i$ for $1 \leq i \leq n$ by a function from $X_i$ to $Y_i$. Then we define a function $H_1 \times \ldots \times H_n$ from $X_1 \times \ldots \times X_n$ to $Y_1 \times \ldots \times Y_n$ by

$(H_1 \times \ldots \times H_n)(x_1, \ldots, x_n) = \langle H_1(x_1), \ldots, H_n(x_n) \rangle.$

It is clear that we can obtain the same information from an oracle for $H_1 \times \ldots \times H_n$ that we can obtain from oracles for each of $H_1, \ldots, H_n$. We therefore define $F$ to be recursive in $H_1, \ldots, H_n$ if $F$ is recursive in $H_1 \times \ldots \times H_n$. In actually describing how to compute $F$, we usually use separate oracles for $H_1, \ldots, H_n$ rather than a single oracle for $H_1 \times \ldots \times H_n$.

## 5. Recursive Enumerability

A set A in X is <u>recursively enumerable</u> (abbreviated <u>RE</u>) if there is a space Y and a recursive set B in X✕Y such that
$$x \in A \leftrightarrow \exists y(\langle x,y \rangle \in B)$$
for all x. In view of the Isomorphism Theorem, we may always take Y = N.

As an example, the equivalence

(1) $\qquad x \in W_I \leftrightarrow \exists n(x \in W_{I,n})$,

together with the fact that $x \in W_{I,n}$ is a recursive relation of x, I, n, shows that $x \in W_I$ is an RE relation of x,I. Similarly,

(2) $\qquad [I](x) = y \leftrightarrow \exists n([I]_n(x) = y)$

shows that $[I](x) = y$ is an RE relation of I, x, y.

A recursive set A is RE; for $x \in A \leftrightarrow \exists y(x \in A)$, and $x \in A$ is a recursive relation of x,y. The converse is false, as the above examples show.

A <u>selector</u> for a relation A in X ✕ Y is a partial function F from X to Y such that for all x, F(x) is defined iff there is a y such that $\langle x,y \rangle \in A$, and, in this case, $\langle x,F(x) \rangle \in A$. Thus F selects a y = F(x) such that $\langle x,y \rangle \in A$, provided that such a y exists.

<u>Selection Theorem</u>. If A is an RE relation in X ✕ Y, then there is a recursive selector for A.

<u>Proof</u>. There is a recursive B such that
$$\langle x,y \rangle \in A \leftrightarrow \exists z(\langle x,y,z \rangle \in B).$$

Let $\{\langle y_n, z_n\rangle\}$ be a listing of $Y \times Z$. We describe an algorithm I from X to Y by describing the process of computing according to I with the input x. The nth step in this process consists of computing $B(x, y_n, z_n)$. If the result is Tr, we give the output $y_n$; otherwise, we go on to the next step. It is clear that [I] is a selector for A.   Q.E.D.

Corollary 1. A set A is RE iff it is the domain of a recursive partial function.

Proof. Such a domain is RE by (1). If A is RE, then $x \in A \leftrightarrow \exists y(\langle x,y\rangle \in B)$ with B recursive. If F is a recursive selector for B, then A is the domain of F.   Q.E.D.

Corollary 2. If B is RE and A is defined by
$$x \in A \leftrightarrow \exists y(\langle x,y\rangle \in B),$$
then A is RE.

Proof. Since A is the domain of any selector for B, this follows from the theorem and Corollary 1.   Q.E.D.

As an example, the range A of a recursive partial function F is RE. For if I is an algorithm for F, then
$$x \in B \leftrightarrow \exists y([I](y) = x)$$
and $[I](y) = x$ is an RE relation of x,y by (2).

By Corollary 1 and the Isomorphism Theorem, a set A in X is RE iff $A = W_I$ for some $I \in \text{Alg}(X,N)$. Any such I is then called an *index* of A.

Parameter Theorem. If A is an RE relation in $X \times Y$, then

there is a recursive function F from Y to Alg(X,N) such that
$$x \in W_{F(y)} \leftrightarrow \langle x,y \rangle \in A$$
for all x and y.

Proof. Let I be an index of A, and let F(y) be the algorithm J such that the process of computing according to J with input x is the same as the process of computing according to I with input $\langle x,y \rangle$. Then F is recursive and
$$x \in W_{F(y)} \leftrightarrow \langle x,y \rangle \in W_I$$
$$\leftrightarrow \langle x,y \rangle \in A. \qquad Q.E.D.$$

Listing Theorem. An infinite set A is RE iff it has a listing.

Proof. If A has a listing, it is the range of a recursive function and hence is RE. Suppose that A is RE. Let $x \in A \leftrightarrow \exists y (\langle x,y \rangle \in B)$ with B recursive, and choose $x_0 \in A$. Set $F(\langle x,y \rangle) = x$ if $\langle x,y \rangle \in B$ and $F(\langle x,y \rangle) = x_0$ otherwise. Then F is recursive and has range A. By the Isomorphism Theorem and Lemma 2 of §2, A has a listing. Q.E.D.

The graph of a partial function F from X to Y is the relation A in X × Y defined by $\langle x,y \rangle \in A \leftrightarrow F(x) = y$.

Graph Theorem. A partial function is recursive iff its graph is RE.

Proof. The graph of [I] is RE by (2). Since the only selector for the graph of F is F itself, the converse follows from the Selection Theorem. Q.E.D

Complementation Theorem. A set A is recursive iff both A

and $A^c$ are RE.

*Proof.* If A is recursive, then $A^c$ is recursive; so A and $A^c$ are RE. Suppose that A and $A^c$ are RE with indices I and J respectively. For each x, $x \in W_I$ or $x \in W_J$; so we may define a recursive function F by

$$F(x) = \mu n(x \in W_{I,n} \lor x \in W_{J,n}).$$

Then $x \in A \leftrightarrow x \in W_{I,F(x)}$; so A is recursive.    Q.E.D.

We now establish some connections between recursiveness and limits. If $\{x_n\}$ is an infinite sequence in X, we say that x is the <u>limit</u> of $\{x_n\}$ and write $\lim x_n = x$ if $x_n = x$ for all sufficiently large n. If $\{F_n\}$ is an infinite sequence of functions from X to Y, we say that F is the <u>limit</u> of $\{F_n\}$ and write $\lim F_n = F$ if $\lim F_n(x) = F(x)$ for all x. If $\lim F_n = F$, a <u>modulus</u> for $\{F_n\}$ is a function H from X to N such that

$$n \geq H(x) \rightarrow F_n(x) = F(x)$$

for all n and x.

A sequence $\{F_n\}$ is <u>recursive</u> if $F_n(x)$ is a recursive function of n,x.

<u>Modulus Lemma</u>. If A is RE and F is recursive in A, then there is a recursive sequence $\{F_n\}$ such that $\lim F_n = F$ and a modulus of $\{F_n\}$ which is recursive in A.

*Proof.* We suppose that all spaces involved are N. Let I be an index of A and let $A_n = W_{I,n}$. Then $\{A_n\}$ is recursive and $\lim A_n = A$. We obtain a modulus G for $\{A_n\}$ by setting

$G(k) = \mu\, n(k \in A_n)$ if $k \in A$ and $G(k) = 0$ if $k \notin A$; and $G$ is recursive in $A$.

Let $J$ be an algorithm for $F$ in $A$. For any $r$, $r \in W_J^A$. Let $H(r)$ be the smallest number $m$ such that $r \in W_{J,m}^A$ and $m \geq G(k)$ for every $k$ used in the computation of $[J]^A(r)$. If $n \geq H(r)$, then $[J]_n^{A_n}(r) = [J]_n^A(r) = [J]^A(r) = F(r)$ by the Use Principle. Thus if we set $F_n(r) = [J]_n^{A_n}(r)$ when $r \in W_{J,n}^{A_n}$ and $F_n(r) = 0$ otherwise, then $\lim F_n = F$ and $H$ is a modulus for $\{F_n\}$. Finally, $H$ is recursive in $A$; for an oracle for $A$ enables us to find the $z$ used in the computation of $[J]^A(r)$ and to compute $G(z)$ for each such $z$. Q.E.D.

A set $A$ in $X$ is RE <u>in</u> $H$ if there is a space $Y$ and a set $B$ in $X \times Y$ recursive in $H$ such that

$$x \in A \leftrightarrow \exists y(\langle x,y\rangle \in B)$$

for all $x$. We can then relativize all of the above results to $H$. An index of $A$ <u>in</u> $H$ is an $I \in \text{Alg}(X,N)$ such that $A = W_I^H$.

If $G$ is recursive in $H$, then any set RE in $G$ is RE in $H$ (because of the corresponding result with <u>recursive</u> in place of <u>RE</u>).

The <u>jump</u> of a function $H$ is the relation $H^*$ in $\text{Alg}(N,N) \times N$ defined by

$$\langle I,n\rangle \in H^* \leftrightarrow n \in W_I^H.$$

Then $H^*$ is RE in $H$. Moreover, every set RE in $H$ is recursive in $H^*$. It suffices to prove this for a set $A$ in $N$. Let $I$

be an index of A in H. Then $n \in A \leftrightarrow \langle I,n \rangle \in H^*$, proving that A is recursive in $H^*$.

Since H is recursive in H and hence RE in H, H is recursive in $H^*$. However, $H^*$ is not recursive in H. If it were, then every set RE in H would be recursive in H, contradicting the relativization to H of the previously proved result that not every RE set is recursive.

## 6. Degrees

We write $F \leq_R G$ if $F$ is recursive in $G$. As we saw in §4,

(1) $\quad F \leq_R F$,

(2) $\quad F \leq_R G \ \& \ G \leq_R H \to F \leq_R H$.

We let $F \sim G$ mean that $F \leq_R G$ and $G \leq_R F$. From (1) and (2) we see that $\sim$ is an equivalence relation among functions. The equivalence class of $F$ is called the _degree_ of $F$, and is designated by $\text{dg } F$. Then by the fundamental property of equivalence classes,

(3) $\quad F \sim G \leftrightarrow \text{dg } F = \text{dg } G$.

We use a, b, c, and d for degrees. (In print, boldface letters are generally used.)

We let $a \leq b$ mean that for some $F$ and $G$, we have $a = \text{dg } F$, $b = \text{dg } G$, and $F \leq_R G$. Then

(4) $\quad \text{dg } F \leq \text{dg } G \leftrightarrow F \leq_R G$.

The implication from right to left is evident. If $\text{dg } F \leq \text{dg } G$, then we have $\text{dg } F = \text{dg } F'$, $\text{dg } G = \text{dg } G'$, and $F' \leq_R G'$. Using (3), $F \leq_R F'$, $F' \leq_R G'$, and $G' \leq_R G$. Hence $F \leq_R G$ by (2).

From (1) through (4) it follows that $\leq$ is a partial ordering of the set of degrees. As usual, we write $a < b$ for $a \leq b$ a $\neq$ b. Using (4) and (3),

(5) $\quad \text{dg } F < \text{dg } G \leftrightarrow F \leq_R G \ \& \ G \not\leq_R F$.

We can think of $F \leq_R G$ as meaning that $F$ is at least as easy to compute as $G$. Then $F \sim G$ means that $F$ and $G$ are equally easy to compute. Thus the degree of $F$ is a measure of the difficulty of computing F; the higher this degree (in the partially ordered set of degrees), the more difficult F is to compute.

If A is the graph of F, then dg F = dg A; for A is recursive in F by the definition of A, and F is recursive in A by the Graph Theorem relativized to A. Thus every degree is the degree of some set. By the Isomorphism Theorem, we see that for any X and Y, every degree is the degree of a function from X to Y and of a set in X.

If F is recursive, then $F \leq_R G$ for all G; so dg $F \leq$ a for all a. Thus there is a smallest degree 0, and 0 is the degree of every recursive function. Conversely, if dg F = 0, then F is recursive. For taking any recursive G, we have $F \leq_R G$; so F is recursive.

We note that the notion of a least upper bound (abbreviated lub) and a greatest lower bound (abbreviated glb) makes sense in any partially ordered set.

Let $a_1, \ldots, a_n$ be degrees, and let $F_1, \ldots, F_n$ be functions such that dg $F_i = a_i$. Using any oracle, we can compute $F_1 \times \ldots \times F_n$ iff we can compute each of $F_1, \ldots, F_n$. In terms of degrees, this means that $dg(F_1 \times \ldots \times F_n) \leq a$ iff a is an upper bound of $\{a_1, \ldots, a_n\}$. Thus dg $F_1 \times \ldots \times F_n$ is the lub of $\{a_1, \ldots, a_n\}$. This lub is designated by $a_1 \cup \ldots \cup a_n$. Note that dg $G \leq a_1 \cup \ldots \cup a_n$ iff G is recursive in $F_1, \ldots, F_n$; and $a_1 \cup \ldots \cup a_n \leq$ dg G iff each $F_i$ is recursive in G.

Since the empty set of degrees has the lub 0, every finite set of degrees has a lub.

If $\{F_n\}$ is an infinite sequence of functions from X to Y, we sometimes identify $\{F_n\}$ with the function F from $N \times X$ to Y defined by $F(n,x) = F_n(x)$. This explains such notation as $\{F_n\} \leq_R G$ or dg $\{F_n\}$.

Clearly $F_n \leq_R \{F_n\}$; so dg $F_n \leq$ dg $\{F_n\}$. This implies that every countable set of degrees has an upper bound.

We write $A \leq_{RE} F$ to mean that A is RE in F. From the results of §5,

(6) $\quad A \leq_R F \rightarrow A \leq_{RE} F$,

(7) $\quad A \leq_{RE} F \ \& \ F \leq_R G \rightarrow A \leq_{RE} G$.

We say that a is <u>RE in</u> b, and write a $\leq_{RE}$ b, if for some A and F we have a = dg A, b = dg F, and $A \leq_{RE} F$. We then have $A \leq_{RE} G$ for every G with dg G = b by (7). However, we do not necessarily have $B \leq_{RE} F$ for every B with dg B = a. For example, dg $A^c$ = dg A = a; but by the relativized Complementation Theorem, we do not have $A^c \leq_{RE} F$ unless $A \leq_R F$.

By (6) and (7),

(8) $\quad a \leq b \rightarrow a \leq_{RE} b$,

(9) $\quad a \leq_{RE} b \ \& \ b \leq c \rightarrow a \leq_{RE} c$.

If a is RE in 0, we say simply that a is RE. Thus a is RE iff it is the degree of an RE set.

For any H, the basic properties of the jump imply that dg H* is the largest degree which is RE in dg H. Thus for any a, there is a largest degree which is RE in a. This degree is called the _jump_ of a, and is designated by a'. As we have just seen, (dg H)' = dg H*. Since H $\leq_R$ H* and H* $\not\leq_R$ H, we have

(11)    a < a'

by (5). Also

(12) a $\leq$ b $\rightarrow$ a' $\leq$ b'.

For a' $\leq_{RE}$ a; so if a $\leq$ b, then a' $\leq_{RE}$ b and hence a' $\leq$ b'.

_Limit Lemma_. A function F is of degree $\leq$ a' iff there is a sequence $\{F_n\}$ such that dg $\{F_n\}$ $\leq$ a and lim $F_n$ = F.

_Proof_. Suppose that such a sequence exists. Define A by

$$\langle x,n \rangle \in A \leftrightarrow (\forall m \geq n)(F_m(x) = F_n(x)).$$

Then

$$\langle x,n \rangle \in A^c \leftrightarrow \exists m(m \geq n \ \& \ F_m(x) \neq F_n(x)).$$

Hence $A^c$ is RE in $\{F_n\}$; so dg A = dg $A^c$ $\leq$ a'. Setting H(x) = $\mu$n($\langle x,n \rangle \in A$), H is a function recursive in A and hence of degree $\leq$ a'. Since F(x) = $F_{H(x)}(x)$, F is recursive in $\{F_n\}$, H, and hence is of degree $\leq$ a'.

Now let dg F $\leq$ a'. Let A be of degree a and let B = A*. Then the Modulus Lemma relativized to A supplies the desired sequence.    Q.E.D.

The Limit Lemma also holds with F and $F_n$ replaced by A and $A_n$. We have only to show that if A = lim $F_n$, then A = lim $A_n$ for some sequence $\{A_n\}$ with $\{A_n\} \leq_R \{F_n\}$. For this it suffices to set $A_n(x)$ = min ($F_n(x)$, 1).

We conclude with some remarks about cardinalities. Since Alg(N,N) has a listing, it is countable. Thus for any fixed F, there are only countably many $[I]^F$ with $I \in$ Alg(N,N), and hence only countably many functions from N to N which are recursive in F. It follows that each degree is the degree of only countably many functions from N to N. Since there are continuum many functions from N to N, there are continuum many degrees. Our result also shows that there are only countably many degrees less than any fixed degree; so no uncountable set of degrees has an upper bound.

The theory of degrees is the theory of the partially ordered set of degrees with the jump operation. We have seen that this set has some nice properties. It has a smallest element, and every finite set has a lub. The jump operator increases degrees, and it preserves $\leq$. The content of much of what follows is that there are very few more nice properties. Before turning to such results, however, we shall develop some methods for evaluating degrees of specific functions and sets.

## 7. Evaluating Degrees

We define the $\Sigma_n$ and $\Pi_n$ sets by induction on n. The $\Sigma_0$ sets and the $\Pi_0$ sets are the recursive sets. A set A is $\Sigma_{n+1}$ if it has a definition
$$x \in A \leftrightarrow \exists y (\langle x,y \rangle \in B)$$
where B is $\Pi_n$. A set A is $\Pi_{n+1}$ if it has a definition
$$x \in A \leftrightarrow \forall y (\langle x,y \rangle \in B)$$
where B is $\Sigma_n$. (In the literature, $\Sigma_n$ and $\Pi_n$ are generally written $\Sigma_n^0$ and $\Pi_n^0$ to distinguish them from other kinds of sets which we do not consider.) Note that the $\Sigma_1$ sets are just the RE sets.

We now give some rules for showing that sets are $\Sigma_n$ or $\Pi_n$. In each case, the proof is by induction on n. The case n = 0 is omitted in the proof, since it follows from previous results.

(A1) If A is $\Sigma_n$ ($\Pi_n$) and B is defined by $x \in B \leftrightarrow F(x) \in A$ where F is recursive, then B is $\Sigma_n$ ($\Pi_n$).

*Proof.* Trivial.

(A2) If A is $\Sigma_m$ or $\Pi_m$ for some m < n, then A is $\Sigma_n$ and $\Pi_n$.

*Proof.* We suppose that A is $\Sigma_m$; the $\Pi_m$ case is similar. By the induction hypothesis, A is $\Sigma_{n-1}$. Now $x \in A \leftrightarrow \forall y (x \in A)$; and by (A1), $x \in A$ is a $\Sigma_{n-1}$ relation of x,y. Hence A is $\Pi_n$. If n = 1, A is also $\Pi_{n-1}$; then a similar proof shows that A is $\Sigma_n$. If n > 1, $x \in A \leftrightarrow \exists y (\langle x,y \rangle \in B)$ where B is $\Pi_{n-2}$. By the induction hypothesis,

B is $\Pi_{n-1}$; so A is $\Sigma_n$.

(A3) If A is $\Sigma_n$ ($\Pi_n$), then $A^c$ is $\Pi_n$ ($\Sigma_n$).

Proof. Say A is $\Sigma_n$. Then $x \in A \leftrightarrow \exists y(\langle x,y \rangle \in B)$ where B is $\Pi_{n-1}$. Hence $x \in A^c \leftrightarrow \forall y(\langle x,y \rangle \in B^c)$. By the induction hypothesis, $B^c$ is $\Sigma_{n-1}$; so $A^c$ is $\Pi_n$.

(A4) If A is $\Sigma_n$ and B is defined by $x \in B \leftrightarrow \exists y(\langle x,y \rangle \in A)$, then B is $\Sigma_n$. If A is $\Pi_n$ and B is defined by $x \in B \leftrightarrow \forall y(\langle x,y \rangle \in B)$, then B is $\Pi_n$.

Proof. Say A is $\Sigma_n$, so that

$$\langle x,y \rangle \in A \leftrightarrow \exists z(\langle x,y,z \rangle \in C)$$

where C is $\Pi_{n-1}$. Then

$$x \in B \leftrightarrow \exists y \exists z(\langle x,y,z \rangle \in C)$$
$$\leftrightarrow \exists w(\langle x,F(w),G(w) \rangle \in C)$$

where w varies through $X \times Y$, $F(x,y) = x$, and $G(x,y) = y$. By (A1), $\langle x,F(w),G(w) \rangle \in C$ is a $\Pi_{n-1}$ relation of x,w; so B is $\Sigma_n$.

(A5) If A and B are $\Sigma_n$ ($\Pi_n$), then $A \cup B$ and $A \cap B$ are $\Sigma_n$ ($\Pi_n$).

Proof. Suppose A and B are $\Sigma_n$, so that $x \in A \leftrightarrow \exists y(\langle x,y \rangle \in C)$ and $x \in B \leftrightarrow \exists z(\langle x,z \rangle \in D)$ where C and D are $\Pi_{n-1}$. Then

$$x \in A \cup B \leftrightarrow \exists y \exists z(\langle x,y \rangle \in C \vee \langle x,z \rangle \in D).$$

By (A1) and the induction hypothesis, the part following $\exists z$ is a $\Pi_{n-1}$ relation of x, y, z. Then the part following

$\exists y$ is a $\Sigma_n$ relation of x,y; so $A \cup B$ is $\Sigma_n$ by (A4). We treat $A \cap B$ similarly.

(A6) If A is $\Sigma_n$ ($\Pi_n$), and B and C are defined by
$$\langle x,k \rangle \in B \leftrightarrow (\forall n < k)(\langle x,n,k \rangle \in A),$$
$$\langle x,k \rangle \in C \leftrightarrow (\exists n < k)(\langle x,n,k \rangle \in A),$$
then B and C are $\Sigma_n$ ($\Pi_n$).

Proof. If A is $\Pi_n$, we have
$$\langle x,k \rangle \in B \leftrightarrow \forall n(k \leq n \vee \langle x,n,k \rangle \in A);$$
using (A2), (A5), and (A4), we conclude that B is $\Pi_n$. If A is $\Sigma_n$, then
$$\langle x,n,k \rangle \in A \leftrightarrow \exists y(\langle x,n,k,y \rangle \in D)$$
where D is $\Pi_{n-1}$. Then
$$\langle x,k \rangle \in B \leftrightarrow (\forall n < k)\exists y(\langle x,n,k,y \rangle \in D)$$
$$\leftrightarrow \exists \sigma(\forall n < k)(\langle x,n,k,\sigma(n) \rangle \in D),$$
where $\sigma$ varies through $Sq(Y)$ and $\sigma(n)$ is taken to be some fixed element of Y if n is not in the domain of $\sigma$. By (A1) and the induction hypothesis, the part following $\exists \sigma$ is $\Pi_{n-1}$; so B is $\Sigma_n$.

To treat C, note that
$$\langle x,k \rangle \in C^c \leftrightarrow (\forall n < k)(\langle x,n,k \rangle \in A^c).$$
If A is $\Sigma_n$ ($\Pi_n$), then $A^c$ is $\Pi_n$ ($\Sigma_n$) by (A3); so $C^c$ is $\Pi_n$ ($\Sigma_n$) by the above; so C is $\Sigma_n$ ($\Pi_n$) by (A3).

We note that (A6) also applies to the quantifiers $(\exists n \leq k)$ and $(\forall n \leq k)$; for we can rewrite then as $(\exists n < k+1)$ and

$(\forall n < k+1)$.

We say that a statement P is $\Sigma_n$ ($\Pi_n$) if it is a $\Sigma_n$ ($\Pi_n$) relation of all the (free) variables in P. Then (A3) tells us that if P is $\Sigma_n$ ($\Pi_n$), then $\neg$P is $\Pi_n$ ($\Sigma_n$). By (A5), if P and Q are $\Sigma_n$ ($\Pi_n$), then P $\vee$ Q and P & Q are $\Sigma_n$ ($\Pi_n$). We can then treat P $\to$ Q and P $\leftrightarrow$ Q by noting that P $\to$ Q is equivalent to $\neg$P $\vee$ Q and P $\leftrightarrow$ Q is equivalent to (P $\to$ Q) & (Q $\to$ P). It is clear that if P is $\Sigma_n$, then $\forall$xP is $\Pi_{n+1}$, and if P is $\Pi_n$, then $\exists$xP is $\Sigma_{n+1}$. By (A4), if P is $\Sigma_n$, then $\exists$xP is $\Sigma_n$, and if P is $\Pi_n$, then $\forall$xP is $\Pi_n$. By (A6), if P is $\Sigma_n$ ($\Pi_n$), then ($\forall n < k$)P and ($\exists n < k$)P are $\Sigma_n$ ($\Pi_n$). We frequently make tacit use of these rules to verify that a set which we have defined is $\Sigma_n$ or $\Pi_n$. Note that we may also use these rules to show that a set is RE by showing that it is $\Sigma_1$.

<u>Post's Theorem</u>. A set A is $\Sigma_{n+1}$ iff it is RE in a $\Pi_n$ set.

<u>Proof</u>. If A is $\Sigma_{n+1}$, then x $\in$ A $\leftrightarrow$ $\exists$y($\langle$x,y$\rangle \in$ B) where B is $\Pi_n$; and A is RE in B. Now suppose that A is RE in the $\Pi_n$ set B in N. Let I be an index of A in B. By (3) of §4,

$$x \in A \leftrightarrow x \in W_I^B$$
$$\leftrightarrow \exists \alpha (\alpha \subset B \ \& \ x \in W_I^\alpha).$$

To show that A is $\Sigma_{n+1}$, it suffices by the rules to show that $\alpha \subset$ B and x $\in W_I^\alpha$ are $\Sigma_{n+1}$. The latter is RE, hence $\Sigma_1$,

hence $\Sigma_{n+1}$. To treat the former, note that
$$\alpha \subset B \leftrightarrow (\forall n < \mathrm{lh}(\alpha))(\alpha(n) = B(n));$$
so by the rules, it suffices to verify that $k = B(n)$ is $\Sigma_{n+1}$. But
$$k = B(n) \leftrightarrow (k = \mathrm{Tr} \ \& \ n \in B) \lor (k = \mathrm{Fa} \ \& \ \neg(n \in B));$$
so $k = B(n)$ is $\Sigma_{n+1}$ by the rules.   Q.E.D.

Corollary. A set A is both $\Sigma_{n+1}$ and $\Pi_{n+1}$ iff it is recursive in a $\Pi_n$ relation.

Proof. Use the relativized Complementation Theorem.  Q.E.D.

We note that by (A3), $\Pi_n$ may be replaced by $\Sigma_n$ in both the theorem and the corollary.

We now connect the above with degrees. We write $a^n$ for a followed by n primes.

Lemma. The highest degree of a $\Sigma_n$ ($\Pi_n$) set is $0^n$.

Proof. We use induction on n. The case $n = 0$ is trivial. If the theorem is true for $\Pi_n$ sets, it is true for $\Sigma_{n+1}$ sets by Post's Theorem and for $\Pi_{n+1}$ sets by (A3). Q.E.D.

This lemma is the principal tool for evaluating the degree of specific sets. We have already seen how to show that a set is $\Sigma_n$ or $\Pi_n$; we now explain how to show that it has the highest degree among such sets.

We say that A is many-one reducible, or simply reducible, to B if there is a recursive function F such that $x \in A \leftrightarrow F(x) \in B$ for all x; i.e., such that $F(A) \subset B$ and $F(A^c) \subset B^c$.

This clearly implies that $A \leq_R B$.

We say that $A$ is <u>complete</u> $\Sigma_n$ (<u>complete</u> $\Pi_n$) if $A$ is $\Sigma_n$ ($\Pi_n$) and every $\Sigma_n$ ($\Pi_n$) set is reducible to $A$. (We say <u>complete RE</u> for complete $\Sigma_1$.) Then any complete $\Sigma_n$ or complete $\Pi_n$ set has degree $0^n$ by the lemma.

It is easy to see that if $A$ is $\Sigma_n$ ($\Pi_n$) and some complete $\Sigma_n$ ($\Pi_n$) set is reducible to $A$, then $A$ is complete $\Sigma_n$ ($\Pi_n$). It is therefore useful to have a supply of complete $\Sigma_n$ and $\Pi_n$ sets. We shall find a few such sets.

Let Tot be the set of $I$ in $Alg(N,N)$ such that $W_I = N$. (Thus $I \in$ Tot iff $[I]$ is total.) We shall show that Tot is complete $\Pi_2$. We have
$$I \in \text{Tot} \leftrightarrow \forall n(n \in W_I);$$
since $n \in W_I$ is RE and hence $\Sigma_1$, Tot is $\Pi_2$. If $A$ is any $\Pi_2$ set, we have $x \in A \leftrightarrow \forall n(\langle x,n\rangle \in B)$ where $B$ is $\Sigma_1$ and hence RE. By the Parameter Theorem, there is a recursive $F$ such that $n \in W_{F(x)} \leftrightarrow \langle x,n\rangle \in B$. Then
$$x \in A \leftrightarrow W_{F(x)} = N$$
$$\leftrightarrow F(x) \in \text{Tot};$$
so $A$ is reducible to Tot.

We say that $A$ is reducible to $B,C$ if there is a recursive function $F$ such that $F(A) \subset B$ and $F(A^c) \subset C$. This implies that $A$ is reducible to any set satisfying $B \subset D$ and $C \cap D = \emptyset$.

Let Fin be the set of $I$ in $Alg(N,N)$ such that $W_I$ is finite

We show that every $\Pi_2$ set is reducible to Tot, Fin. By the above, it suffices to show that Tot is reducible to Tot, Fin. By the rules of this section, $(\forall k \leq n)(k \in W_I)$ is an RE relation of $n, I$; so by the Parameter Theorem, there is a recursive function F such that
$$n \in W_{F(I)} \leftrightarrow (\forall k \leq n)(k \in W_I).$$
It readily follows that $F(\text{Tot}) \subset \text{Tot}$ and $F(\text{Tot}^c) \subset \text{Fin}$.

A consequence of this result is that every $\Pi_2$ set is reducible to $\text{Fin}^c$. Now
$$I \in \text{Fin}^c \leftrightarrow \forall n \exists k (n < k \ \& \ k \in W_I);$$
so $\text{Fin}^c$ is $\Pi_2$. Thus $\text{Fin}^c$ is complete $\Pi_2$; so Fin is complete $\Sigma_2$.

Now let Cof be the set of I in Alg(N,N) such that $W_I$ is cofinite. We show that every $\Pi_2$ set is reducible to Tot, Cof - Tot. In view of the above, it suffices to find a recursive F such that $F(\text{Tot}) \subset \text{Tot}$ and $F(\text{Fin}) \subset \text{Cof} - \text{Tot}$.

Choose F by the Parameter Theorem so that
$$n \in W_{F(I)} \leftrightarrow (\forall r \leq n)(r \in W_I) \vee [(\exists k < n)(k \notin W_{I,n})$$
$$\& \ (\forall k < n)(k \in W_{I,n+1} \to k \in W_{I,n})].$$
Clearly $F(\text{Tot}) \subset \text{Tot}$. Now let $W_I$ be finite; we must show that $W_{F(I)}$ is cofinite and different from N. If n is sufficiently large, $(\exists k < n)(k \notin W_I)$ and $W_{I,n} = W_{I,n+1} = W_I$. All such n are in $W_{F(I)}$; so $W_{F(I)}$ is cofinite. Let r be the smallest number not in $W_I$. Either $r \notin W_{F(I)}$, or there is a $k < r$

such that $k \notin W_{I,r}$. In the latter case, $k \in W_I$; so there is an n such that $n \geq r$, $k \notin W_{I,n}$, and $k \in W_{I,n+1}$. Since $k < n$, we have $n \notin W_{F(I)}$. Thus $W_{F(I)} \neq N$.

We now show that Cof is complete $\Sigma_3$. Since
$$I \in \text{Cof} \leftrightarrow \exists n \forall k (k \leq n \vee k \in W_I),$$
Cof is $\Sigma_3$. Now let A be $\Sigma_3$. Then $x \in A \leftrightarrow \exists n (\langle x,n \rangle \in B)$ where B is $\Pi_2$. Then $(\exists n < k)(\langle x,n \rangle \in B)$ is a $\Pi_2$ relation of $x,n$; so by the above, there is a recursive F such that
$$(\exists n < k)(\langle x,n \rangle \in B) \to F(x,k) \in \text{Tot},$$
$$\neg (\exists n < k)(\langle x,n \rangle \in B) \to F(x,k) \in \text{Cof} - \text{Tot}.$$
Choose G by the Parameter Theorem so that G is recursive and
$$\langle k,m \rangle \in W_{G(x)} \leftrightarrow m \in W_{F(x,k)}.$$
If $x \in A$, then for all sufficiently large k we have $F(x,k) \in \text{Tot}$. Since $F(x,k) \in \text{Cof}$ for all k, it follows that $W_{G(x)}$ is cofinite. If $x \notin A$, then $F(x,k) \notin \text{Tot}$ for all k; it follows that $W_{G(x)}$ is coinfinite. The Isomorphism Theorem now shows that A is reducible to Cof.

The results of this section can be relativized to any H. In the definition of $\Sigma_n$ in H and $\Pi_n$ in H, which we also write as $\Sigma_n[H]$ and $\Pi_n[H]$, the only difference is that the sets $\Sigma_0$ in H or $\Pi_0$ in H are now the sets recursive in H. Our list of rules and Post's Theorem are relativized by replacing $\Sigma_n$ and $\Pi_n$ by $\Sigma_n[H]$ and $\Pi_n[H]$. The relativized lemma states that the highest degree of $\Sigma_n[H]$ or $\Pi_n[H]$ sets is $(\text{dg } H)^n$.

## 8. Incomparable Degrees

The results of the last section may suggest that the degrees $0^n$ are the only small degrees. We shall show that, on the contrary, there are degrees between $0$ and $0'$.

We say that the degrees a and b are <u>incomparable</u>, and write a|b, if neither $a \leq b$ nor $b \leq a$.

<u>Theorem</u> (Kleene-Post). There are degrees a and b such that a|b, $a \leq 0'$, and $b \leq 0'$.

<u>Corollary</u>. There is a degree a such that $0 < a < 0'$.

<u>Proof</u>. With a and b as in the theorem, $0 \leq a \leq 0'$. Since $0 \leq b \leq 0'$ and a|b, we have $a \neq 0$ and $a \neq 0'$. Q.E.D.

Since the proof of the theorem will be a model for several later proofs, we give it in considerable detail.

Let F and G be partial functions from X to Y. We say that F and G are <u>compatible</u>, and write Comp(F,G), if F and G have a common extension; otherwise, F and G are <u>incompatible</u>. It is easy to see that F and G are incompatible iff there is an x such that F(x) and G(x) are defined and distinct.

We shall first concentrate on obtaining incomparable degrees a and b. Since it is difficult to construct degrees directly, we shall construct functions F and G from N to N and take $a = \text{dg } F$, $b = \text{dg } G$. Then we want to have $F \not\leq_R G$ and $G \not\leq_R F$. This may be broken down into the infinite set of conditions (where I varies through Alg(N,N)):

$$(1_I) \quad F \neq [I]^G;$$

$(2_I)$ $G \neq [I]^F$.

Each condition is a finite object (viz., an expression); and the class of conditions is a space. We let $\{R_s\}$ be a listing of the space of conditions.

We shall construct F and G in steps. At each step, we define finitely many values of F and G. At step s, we <u>insure</u> condition $R_s$; i.e., we define some values of F and G so that no matter how the remaining values are defined, $R_s$ will hold. At the end of all the steps, F and G will be completely defined and will satisfy all the conditions.

We let $F^S$ and $G^S$ be the finite parts of F and G defined before step s; i.e., $F^S(k) = n$ iff we have set $F(k) = n$ at some step before step s.

We now describe step s. We suppose that $R_s$ is $(1_I)$; if it is $(2_I)$, we merely interchange F and G. Let n be the smallest number such that $F^S(n)$ is undefined. There are two cases.

Case 1: There is a finite function $\sigma$ such that $\text{Comp}(\sigma, G^S)$ and $n \in W_I^\sigma$.

We then choose such a $\sigma$. We set $G(m) = \sigma(m)$ for each m in the domain of $\sigma$ for which $G^S(m)$ is undefined; and we set $F(n) = [I]^\sigma(n) + 1$. Since $\text{Comp}(\sigma, G^S)$, these choices guarantee that $\sigma \subset G$. Hence we will have $[I]^G(n) = [I]^\sigma(n) \neq F(n)$; so $(1_I)$ will hold.

Case 2: Otherwise.

We set $F(n) = 0$. To show that $(1_I)$ will hold, it will suffice to show that we will have $n \notin W_I^G$. Suppose that $n \in W_I^G$. By (1) of §4, there is a $\sigma \subset G$ such that $n \in W_I^\sigma$. Since $\sigma$ and $G^S$ have the common extension $G$, $\text{Comp}(\sigma, G^S)$. This contradicts the case hypothesis.

At infinitely many steps we have defined $F(n)$, where $n$ was the smallest number for which $F(n)$ was previously undefined. It follows that $F(n)$ is defined for all $n$; and similarly for $G$. We thus have incomparable degrees $\text{dg } F$ and $\text{dg } G$. To complete the proof, we must show that $\text{dg } F \leq 0'$ and $\text{dg } G \leq 0'$.

We can identify the construction with a function whose value at $s$ is a description of which values of $F$ and $G$ were assigned at step $s$. Since each value of $F$ and $G$ is assigned at some step, $F$ and $G$ are recursive in the construction. Thus we need only show that the construction has degree $\leq 0'$. In other words, we must show that given an oracle for a suitable function of degree $\leq 0'$, we can, given $s$, compute what happens at step $s$.

We may suppose that we have already computed what happened at all previous steps, so that we know $F^S$ and $G^S$. Since $\{R_s\}$ is a listing, we can find $R_s$. We can also compute the $n$ of step $s$. Now Case 1 holds iff

(1) $\quad \exists \sigma (\text{Comp}(\sigma, F^S) \ \& \ n \in W_I^\sigma)$.

Since $\text{Comp}(\sigma,\pi)$ is a recursive relation of $\sigma$, $\pi$, (1) is an RE relation of $F^S$, I, n (by the rules of the last section). This relation is then of degree $\leq 0'$. Given an oracle for this relation, we can decide which case holds.

If we are in Case 2, there is no problem. If we are in Case 1, the only remaining problem is to find $\sigma$. Now the part after $\exists \sigma$ in (1) is an RE relation of $\sigma$, $F^S$, I, n. Hence by the Selection Theorem, there is a recursive partial function L such that whenever we are in Case 1, $L(F^S, I, n)$ is a suitable $\sigma$. Then we may compute this $\sigma$ without use of an oracle. Q.E.D.

Let us consider the result of relativizing the above proof to a function H. This means that an oracle for H is available at all times. Hence when we do recursion in G, we have oracles for G and H, or, equivalently, and oracle for G $\times$ H. Thus the F which we obtains satisfies $F \not\leq_R G \times H$. Similarly, $G \not\leq_R F \times H$. Taking $a = \text{dg}(F \times H) = \text{dg } F \cup \text{dg } H$ and $b = \text{dg}(G \times H) = \text{dg } G \cup \text{dg } H$, we see that $a|b$. Since (1) is now RE in H, we need an oracle for a function of degree $\leq (\text{dg } H)'$. Thus setting $c = \text{dg } H$, we have $a \leq c'$ and $b \leq c'$. Since clearly $c \leq a$ and $c \leq b$, we have:

Relativized Theorem. For any degree c, there are degrees a and b such that $a|b$, $c \leq a \leq c'$, and $c \leq b \leq c'$.

There are similar relativized results for all of the results which we prove about degrees; but we shall not mention them unless they are of special interest.

## 9. Upper and Lower Bounds

We have seen that every finite set of degrees has a lub. We shall now see that this is false for glb's, and that very simple infinite sets of degrees may fail to have a lub.

An <u>ascending</u> sequence of degrees is an infinite sequence $\{a_n\}$ such that $a_n < a_{n+1}$ for all n. For example, $0, 0', 0'', \ldots$ is an ascending sequence.

<u>Theorem</u> (Kleene-Post-Spector). Let $\{a_n\}$ be an ascending sequence of degrees. Then there are upper bounds b and c for $\{a_n\}$ such that no upper bound of $\{a_n\}$ is a lower bound of $\{b,c\}$.

<u>Corollary 1</u>. No ascending sequence of degrees has a lub.

<u>Corollary 2</u>. There are degrees b and c such that $\{b,c\}$ has no glb.

<u>Proof</u>. Take any ascending sequence $\{a_n\}$, and let b and c be as in the theorem. Then a glb of $\{b,c\}$ would be an upper bound of $\{a_n\}$. Q.E.D.

Now we turn to the proof of the theorem. Let $H_n$ be a function from N to N of degree $a_n$. We construct functions F and G from N × N to N and take $b = dg\ F$, $c = dg\ G$. To insure that no upper bound of $\{a_n\}$ is a lower bound of $\{b,c\}$, we must satisfy the conditions:

$(1_{I,J})$ If $[I]^F = [J]^G = L$, then $L \leq_R H_n$ for some n.

Let $\{R_s\}$ be a listing of the space of conditions. Again we will insure $R_s$ at step s. At step s, we will first define finitely many values of F and G; we will then set

(1)  $F(s,k) = H_s(k)$ if $F(s,k)$ is undefined,
(2)  $G(s,k) = H_s(k)$ if $G(s,k)$ is undefined.

This will first of all insure that F and G are completely defined by the construction. Moreover, $F(s,k)$ can be defined for only finitely many values of k before the last part of step s; so $F(s,k) = H_s(k)$ for all but finitely many k. This shows that $H_s \leq_R F$; so b is an upper bound of $\{a_n\}$. Similarly, c is an upper bound of $\{a_n\}$.

We now describe step s. Let $R_s$ be $(1_{I,J})$, and let $F^S$ and $G^S$ be the parts of F and G already defined.

Case 1: There are $\sigma$ and $\pi$ such that $\text{Comp}(\sigma, F^S)$, $\text{Comp}(\pi, G^S)$, and $[I]^\sigma$ and $[J]^\pi$ are incompatible.

We then choose such a $\sigma$ and $\pi$. We set $F(n,k) = \sigma(n,k)$ for all $(n,k)$ in the domain of $\sigma$ such that $F^S(n,k)$ is undefined; and we set $G(n,k) = \pi(n,k)$ for all $(n,k)$ in the domain of $\pi$ such that $G^S(n,k)$ is undefined. We then proceed to (1) and (2). We have insured that $\sigma \subset F$ and $\pi \subset G$. Hence $[I]^\sigma \subset [I]^F$ and $[J]^\pi \subset [J]^G$; so $[I]^F$ and $[J]^G$ are incompatible. This implies that $(1_{I,J})$ holds.

Case 2. Otherwise.

We then proceed immediately to (1) and (2). We must show that $(1_{I,J})$ holds.

We first show that the set of $\sigma$ such that $\text{Comp}(\sigma, F^S)$ is recursive in $H_s$. We make a list of the finite number of values of $F^S$ not assigned by (1) and (2). To determine whether or not

$Comp(\sigma, F^S)$, it will suffice to determine for each $(n,k)$ in the domain of $\sigma$ whether or not $(n,k)$ is in the domain of $F^S$, and, if it is, determine $F^S(n,k)$. If $F^S(n,k)$ is in out list, we can do this without an oracle. Otherwise, $(n,k)$ is in the domain of $F^S$ iff $n < s$; and in this case, $F^S(n,k) = H_n(k)$. Now for $n < s$, $a_n < a_s$ and hence $H_n \leq_R H_s$. Hence an oracle for $H_s$ will suffice.

Now assume that $[I]^F = [J]^G = L$. We first show that

(3)  $L(i) = j \leftrightarrow \exists \sigma (Comp(\sigma, F^S) \ \& \ [I]^\sigma(i) = j)$.

If $L(i) = j$, then $[I]^F(i) = j$; so by (2) of §4, there is a $\sigma \subset F$ such that $[I]^\sigma(i) = j$. Since $\sigma$ and $F^S$ have the common extension $F$, $Comp(\sigma, F^S)$. Now suppose that $Comp(\sigma, F^S)$ and $[I]^\sigma(i) = j$ for some $\sigma$. Since $[J]^G(i) = L(i)$, there is a $\pi \subset G$ such that $[J]^\pi(i) = L(i)$; and $Comp(\pi, G^S)$. By the case hypothesis, $Comp([I]^\sigma, [J]^\pi)$; so $L(i) = j$.

Now $Comp(\sigma, F^S)$, as a relation of $\sigma$, $i$, $j$, is recursive in $H_s$ and hence RE in $H_s$; while $[I]^\sigma(i) = j$, as a relation of $\sigma$, $i$, $j$, is RE and hence RE in $H_s$. Hence by (3) and the relativizations of the rules of §7, the graph of $L$ is RE in $H_s$. Thus $L \leq_R H_s$ by the relativized Graph Theorem; so $(1_{I,J})$ holds.   Q.E.D.

## 10. The Jump Operation

We are going to determine the range of the jump operation. By (12) of §6, $0' \leq a'$ for all $a$; so every degree in this range is $\geq 0'$. We now show that, conversely, every degree $\geq 0'$ is the jump of some degree.

<u>Theorem</u> (Friedberg). If $0' \leq a$, then there is a degree $b$ such that $b' = b \cup 0' = a$.

While this theorem gives a nice property of the jump operation, it can be used to show that this operation has properties which are not so nice. For example, it is not one-one. For this, use the theorem to obtain a degree $b$ such that $b' = b \cup 0' = 0''$. Then $b' = 0''$, but $b \neq 0'$; for this would imply that $b \cup 0' = 0' \neq 0''$.

In fact, we can have $a' = b'$ with any of the four alternatives $a = b$, $a | b$, $a < b$, $b < a$. The first is obviously possible. For the second, take $a = 0'$ and let $b$ be as in the above paragraph. If $0'$ and $b$ were comparable, we would have either $b \cup 0' = b$ or $b \cup 0' = 0'$; whereas $b \cup 0' = 0''$ which is different from $b$ and $0'$ (since $b \neq b'$). To get the third alternative, we take $a = 0$. We must then choose $b$ so that $b \neq 0$ and $b' = 0'$; and we shall see that this is possible in §13. To get the fourth alternative, we interchange $a$ and $b$ in the third alternative. The third alternative shows that we cannot replace $\leq$ by $<$ in (12) of §6.

Now we turn to the proof of the theorem. For every $b$,

we have $b < b'$ and $0' \leq b'$, so that $b \cup 0' \leq b'$. Thus we have only to construct a $b$ such that $b' \leq a \leq b \cup 0'$.

Let $G$ be a function from $N$ to $N$ of degree $a$. We shall construct a function $F$ from $N$ to $N$ and set $b = \mathrm{dg}\, F$. We will make $F^*$ and $G$ recursive in the construction. This will give an upper bound for $\mathrm{dg}\, F^* = b'$ and $\mathrm{dg}\, G = a$; these upper bounds will suffice to prove $b' \leq a \leq b \cup 0'$. To make $F^*$ recursive in the construction, we shall, for each $I$ and $n$, insure at some step of the construction that $\langle I,n \rangle \in F^*$ or $\langle I,n \rangle \notin F^*$, i.e., that $n \in W_I^F$ or $n \notin W_I^F$. When we do this, we will say that we have <u>decided</u> $n \in W_I^F$. To make $G$ recursive in the construction, we make each $G(n)$ a value of $F$ at some argument.

We are thus led to the following conditions.

($1_{I,n}$) Decide $n \in W_I^F$.

($2_n$) Set $F(m) = G(n)$ for some $m$.

We let $\{R_s\}$ be a listing of the space of conditions.

We now describe step $s$. First suppose that $R_s$ is ($1_{I,n}$).

Case 1: There is a $\sigma$ such that $\mathrm{Comp}(\sigma, F^s)$ and $n \in W_I^\sigma$.

We then pick such a $\sigma$ and, as in previous proofs, assign values of $F$ which guarantee that $\sigma \subset F$. This guarantees that $n \in W_I^F$.

Case 2: Otherwise.

We then assign no new values. We claim that we have insured that $n \notin W_I^F$. For suppose that $n \in W_I^F$. Then there is a $\sigma \subset F$ such that $n \in W_I^\sigma$. Since $\sigma \subset F$, $\mathrm{Comp}(\sigma, F^s)$; and this contradicts the case hypothesis.

Now suppose that $R_s$ is $(2_n)$. We then let m be the smallest number such that $F^s(m)$ is undefined, and set $F(m) = G(n)$.

In each of the infinitely many steps devoted to the conditions $(2_n)$, we have defined F at the smallest argument for which it was previously undefined. Hence F is completely defined at the end of the construction. Moreover, it is clear that $F^*$ and G are recursive in the construction.

Now let us see what oracles are needed to see what happens at step s of the construction. We assume that we already know what happened at previous steps, so that we know $F^s$. We can find $R_s$. If $R_s$ is $(1_{I,n})$, we can show as in the proof in §8 that an oracle for a relation of degree $\leq 0'$ suffices. If $R_s$ is $(2_n)$, we can find m; but we need an oracle to tell us the common value of $F(m)$ and $G(n)$. Thus an oracle for either F or G suffices.

We have shown that the degree of the construction is $\leq b \cup 0'$ and $\leq a \cup 0' = a$. Since $F^*$ and G are recursive in the construction, $b' \leq a$ and $a \leq b \cup 0'$. Q.E.D.

## 11. Minimal Degrees

A degree a is <u>minimal</u> if a > 0 but there is no degree b such that a > b > 0. Thus the corollary in §8 tells us that 0' is not minimal.

<u>Theorem</u> 1 (Spector). There is a minimal degree.

A set A is I-<u>minimal</u> if either $[I]^A$ is not total; or $[I]^A$ is a recursive function; or $[I]^A$ is total and $A \leq_R [I]^A$. Then dg A is minimal iff A is non-recursive and I-minimal for all I. We now develop some methods for obtaining I-minimal sets.

We use St for the space of strings. If i = 0 or i = 1, $\alpha^{(i)}$ is the string obtained from $\alpha$ by adding i at the end. We say $\alpha$ and $\beta$ <u>split</u> $\gamma$ if $\gamma \subset \alpha$, $\gamma \subset \beta$, and $\alpha$ and $\beta$ are incompatible.

A <u>tree</u> is a recursive partial function T from St to St satisfying the condition: if one of $T(\alpha^{(0)})$ and $T(\alpha^{(1)})$ is defined, then all of $T(\alpha)$, $T(\alpha^{(0)})$, and $T(\alpha^{(1)})$ are defined, and $T(\alpha^{(0)})$ and $T(\alpha^{(1)})$ split $T(\alpha)$. For example, the identity mapping from St to St is a tree; it is designated by Id. We use T to designate trees.

A string is <u>on</u> a tree T if it belongs to the range of T. A set A in N is <u>on</u> T if $\alpha \subset A$ for infinitely many $\alpha$ on T.

We say that T' is a <u>subtree</u> of T if every string on T' is on T. This implies that every set on T' is on T.

We say that $\alpha$ and $\beta$ I-<u>split</u> $\gamma$ if $\gamma \subset \alpha$, $\gamma \subset \beta$, and $[I]^\alpha$ and $[I]^\beta$ are incompatible. This implies that $\alpha$ and $\beta$

split $\gamma$; for if $\delta$ were a common extension of $\alpha$ and $\beta$, then $[I]^\delta$ would be a common extension of $[I]^\alpha$ and $[I]^\beta$. We say that $\gamma$ I-<u>splits</u> <u>on</u> T if some $\alpha$ and $\beta$ on T I-split $\gamma$.

<u>Lemma 1</u>. If A is on T, $\alpha \subset A$, and $\alpha$ does not I-split on T, then A is I-minimal.

<u>Proof</u>. We assume that $[I]^A$ is total and prove that it is recursive. For this, we show that

$$[I]^A(m) = k \leftrightarrow \exists \beta (\alpha \subset \beta \ \& \ \beta \text{ on T } \& \ [I]^\beta(m) = k).$$

Since the set of $\beta$ on T is the range of a recursive partial function and hence is RE, this equivalence implies that the graph of $[I]^A$ is RE; so $[I]^A$ is recursive by the Graph Theorem.

Let $[I]^A(m) = k$. Choose $\delta \subset A$ so that $[I]^\delta(m) = k$. For all but finitely many $\beta \subset A$, we have $\alpha \subset \beta$ and $\delta \subset \beta$. Hence there is a $\beta$ on T such that $\alpha \subset \beta \subset A$ and $\delta \subset \beta$. Then $[I]^\beta(m) = k$.

Now let $\alpha \subset \beta$ with $\beta$ on T and $[I]^\beta(m) = k$. By the above, there is a $\gamma$ on T such that $\alpha \subset \gamma$ and $[I]^\gamma(m) = [I]^A(m)$. Since $\beta$ and $\gamma$ do not I-split $\alpha$, $[I]^A(m) = k$. Q.E.D.

A tree T is I-<u>splitting</u> if whenever $T(\alpha^{(0)})$ and $T(\alpha^{(1)})$ are defined, they I-split $T(\alpha)$.

<u>Lemma 2</u>. If T is I-splitting and A is on T, then A is I-minimal.

<u>Proof</u>. We assume that $F = [I]^A$ is total and prove that $A \leq_R F$.

Let B be the set of $\alpha$ such that $T(\alpha)$ is defined and

$T(\alpha) \subset A$. By the definition of a tree,

(1) $\quad \alpha^{(i)} \in B \to \alpha \in B \ \& \ \alpha^{(1-i)} \notin B$.

If $k_n$ is the number of strings of length n in B, then (1) shows that $k_{n+1} \leq k_n$. But B is infinite and clearly $k_0 \leq 1$; so $k_n = 1$ for all n.

Let $G(n)$ be the string of length n in B. We show that $G \leq_R F$. For this, we assume that we have an oracle for F and compute $G(n)$ by induction on n. Clearly $G(n) = 0$. Now suppose that $\alpha = G(n)$ is known. From (1), we see that $G(n+1)$ is either $\alpha^{(0)}$ or $\alpha^{(1)}$; we want to know which. Since $T(G(n+1))$ is defined, $T(\alpha^{(0)}) = \beta$ and $T(\alpha^{(1)}) = \gamma$ are defined and I-split $T(\alpha)$; so for some k and s, $[I]_s^\beta(k)$ and $[I]_s^\gamma(k)$ are defined and distinct. We can compute $\beta$ and $\gamma$; and we can then find such a k and s by running through a listing of $N \times N$. If $G(n+1) = \alpha^{(0)}$, then $\beta \subset A$ and hence $F(k) = [I]_s^\beta(k)$. Similarly, if $G(n+1) = \alpha^{(1)}$, then $F(k) = [I]_s^\gamma(k)$. Our oracle for F which tell us which of these is the case.

To complete the proof, we show that $A \leq_R G$. By the definition of a tree, $\mathrm{lh}(T(G(n+1))) > \mathrm{lh}(T(G(n)))$; so $\mathrm{lh}(T(G(n))) \geq n$. It follows that $A(n) = T(G(n+1))_n$. Q.E.D.

Next we show how to construct I-splitting trees. If J is an algorithm for T, then $\alpha$ and $\beta$ I-split $\gamma$ and are on T iff

$$\gamma \subset \alpha \ \& \ \gamma \subset \beta \ \& \ \exists \delta ([J](\delta) = \alpha) \ \& \ \exists \delta ([J](\delta) = \beta)$$
$$\& \ \exists n \exists i \exists j ([I]^\alpha(n) = i \ \& \ [I]^\beta(n) = j \ \& \ i \neq j).$$

This is an RE relation of I, J, $\alpha$, $\beta$, $\gamma$. Hence by the Selection Theorem, there are recursive partial functions $L_0$ and $L_1$ such that if $\gamma$ I-splits on T, then $L_0(I,J,\gamma)$ and $L_1(I,J,\gamma)$ are on T and I-split $\gamma$; while otherwise, $L_0(I,J,\gamma)$ and $L_1(I,J,\gamma)$ are undefined.

Now let $\delta$ be on T, and define $T'(\gamma)$ by induction on $lh(\gamma)$ as follows:

$$T'(0) = \delta,$$
$$T'(\gamma^{(i)}) = L_i(I,J,T'(\gamma)).$$

It is easy to verify that $T'$ has the following properties: (a) $T'$ is a tree; (b) $T'$ is a subtree of T; (c) $\delta$ is on $T'$; (d) every string on $T'$ is an extension of $\delta$; (e) $T'$ is I-splitting; (f) if $T'(\gamma)$ is defined and I-splits on T, then $T'(\gamma^{(0)})$ and $T'(\gamma^{(1)})$ are defined. We call $T'$ the I-<u>splitting</u> <u>subtree</u> <u>of</u> T <u>for</u> $\delta$.

To prove Theorem 1, we construct a set A in N such that dg A is minimal. Let $\{I_s\}$ be a listing of Alg(N,N). Then we must insure that $A \neq [I_s]$ and that A is $I_s$-minimal for each s.

At step s, we define a total tree $T_s$ and a string $\delta_s$ on $T_s$. Our intention is to take A on $T_s$ so that $\delta_s \subset A$. We insure that $A \neq [I_s]$ and that A is $I_s$-minimal at step s+1. In other words, we will choose $T_{s+1}$ and $\delta_{s+1}$ so that if A is on $T_{s+1}$ and $\delta_{s+1} \subset A$, then $A \neq [I_s]$ and A is $I_s$-minimal.

We must also be sure that we will be able to choose such an A. For this, we choose $T_{s+1}$ and $\delta_{s+1}$ so that $\delta_{s+1}$ is a proper extension of $\delta_s$ and $T_{s+1}$ is a subtree of $T_s$. The former guarantees that there is a unique A such that $\delta_s \subset A$ for all s. Since $\delta_s$ is on $T_s$ for all s, the latter guarantees that $\delta_s$, $\delta_{s+1}$, ... are all on $T_s$; so A is on $T_s$.

We now describe step s. At step 0, we set $T_0$ = Id and $\delta_0 = 0$. Now suppose that step s is completed; we shall do step s+1. Since $\delta_s$ is on $T_s$ and $T_s$ is a total tree, $\delta_s$ has two incompatible extensions on $T_s$. Hence there is a proper extension $\delta$ of $\delta_s$ on $T_s$ such that $\delta$ is incompatible with $[I_s]$. We take $\delta_{s+1}$ to be an extension of $\delta$; this guarantees that $\delta_{s+1}$ is a proper extension of $\delta_s$ and that $A \neq [I_s]$.

There are now two cases. Suppose first that some extension of $\delta$ on $T_s$ does not $I_s$-split on $T_s$. We then let $\delta_{s+1}$ be such an extension, and take $T_{s+1} = T_s$. Then A is $I_s$-minimal by Lemma 1.

Now suppose that there is no such extension. Then we let $\delta_{s+1} = \delta$ and let $T_{s+1}$ be the $I_s$-splitting subtree of $T_s$ for $\delta$. Using (d) and (f) in the properties of splitting subtrees and the fact that every extension of $\delta$ on $T_s$ $I_s$-splits on $T_s$, we easily prove by induction on $lh(\gamma)$ that $T_{s+1}(\gamma)$ is defined. Thus $T_{s+1}$ is a total tree; and A is $I_s$-minimal by Lemma 2. Q.E.D.

One can show that the set we have just constructed has degree $\leq 0''$. Instead, we prove a better result.

Theorem 2 (Sacks). There is a minimal degree a such that $a \leq 0'$.

The construction for Theorem 2 is a modification of that for Theorem 1. Again we define $\delta_s$ at step s so that $\delta_{s+1}$ is a proper extension of $\delta_s$, and take A to be the unique set such that $\delta_s \subset A$ for all s. However, the trees are treated differently. At step s, we define trees $T_i^s$ for $i \leq k_s$ (which need not be total). For $i < k_s$, $T_{i+1}^s$ will be a subtree of $T_i^s$; and $\delta_s$ will be on $T_{k_s}^s$ and hence on all the $T_i^s$. We will always have $T_0^s = \mathrm{Id}$. At the end of the construction, we show that $T_i^s$ converges in a suitable sense to a tree $T_i$.

Now we describe step s. For $s = 0$, set $\delta_0 = 0$, $k_0 = 0$, $T_0^0 = \mathrm{Id}$. Now suppose that step s is completed; we shall do step s+1. Let k be the largest number such that $k \leq k_s$ and $\delta_s$ has a proper extension on $T_k^s$ which is incompatible with $[I_s]$. (Such a k must exists, since $T_0^s = \mathrm{Id}$.) Let $\delta_{s+1}$ be such an extension; this insures that $A \neq [I_s]$. Let $k_{s+1} = k + 1$; and for $i \leq k$, let $T_i^{s+1} = T_i^s$. It remains to choose a subtree $T_{k+1}^{s+1}$ of $T_k^{s+1} = T_k^s$ so that $\delta_{s+1}$ is on $T_{k+1}^{s+1}$.

Case 1: $k = k_s$. Let $T_{k+1}^{s+1}$ be the $I_k$-splitting subtree of $T_k^{s+1}$ for $\delta_{s+1}$.

Case 2: $k < k_s$. Let $T_{k+1}^{s+1} = T_k^{s+1}$.

We prove by induction on i that $k_s = i$ for only finitely many s. Since $k_s = 0$ only for $s = 0$, we may suppose that $i > 0$. Choose $s_0$ by the induction hypothesis so that $k_s \geq i$ for $s \geq s_0$. We may suppose that $k_s = i$ for some $s > s_0$, since otherwise our result is clear. We may therefore choose $s \geq s_0$ so that $k_{s+1} = i$. Then the k at step s+1 is i-1. Since $k_s \geq i > k$, Case 2 occurs at step s+1; so $T_i^{s+1} = T_{i-1}^{s+1}$.

If our result is false, then there is a smallest $t > s$ such that $k_{t+1} = i$. Since $k_{s+2}, \ldots, k_t$ are all $> i$, we have $T_i^{s+1} = T_i^t$ and $T_{i-1}^{s+1} = T_{i-1}^t$; so $T_i^t = T_{i-1}^t$. This implies that the k at step t+1 is not i-1; so $k_{t+1} \neq i$, a contradiction.

If $k_s > i$ for all $s \geq s_0$, then $T_i^s$ is defined and equal to $T_i^{s_0}$ for all $s \geq s_0$. We let $T_i$ be $T_i^{s_0}$. Since $\delta_s$ is on $T_i^s$ whenever $T_i^s$ is defined, $\delta_s$ is on $T_i$; so A is on $T_i$.

We shall show that if $T_{i+1}^s$ is defined, then either it is an $I_i$-splitting subtree of $T_i^s$, or $T_{i+1}^s = T_i^s$ and some $\delta_n$ does not $I_i$-split on $T_i^s$. Assuming this, we see that the same result holds when all of the superscripts s are dropped; and it then follows by Lemmas 1 and 2 that A is $I_i$-minimal.

We prove this result by induction on s. If $s = 0$, $T_{i+1}^s$ cannot be defined. Assume that the result holds for some s. If $k_{s+1} > i+1$, then $T_{i+1}^{s+1} = T_{i+1}^s$ and $T_i^{s+1} = T_i^s$; so this case is trivial. If $k_{s+1} < i+1$, then $T_{i+1}^{s+1}$ is not defined. Hence we may suppose that $k_{s+1} = i+1$, so that the k at step s+1 is i. If Case 1 occurs at step s+1, the result is obvious; so we suppose that Case 2 occurs. Certainly $T_{i+1}^{s+1} = T_i^{s+1}$. Since $i = k < k_s$, $T_{i+1}^s$ is defined; and by the choice of k, no proper extension of $\delta_s$ on $T_{i+1}^s$ is incompatible with $[I_s]$. By the definition of a tree, $\delta_s$ can have no proper extension on $T^s_{i+1}$. The choice of k also implies that $T^s_i \neq T^s_{i+1}$; so $T^s_{i+1}$ is an $I_1$-splitting subtree of $T^s_i$ by the induction hypothesis. By property (f) of splitting subtrees, $\delta_s$ does not $I_1$-split on $T^s_i = T^{s+1}_i$.

The only thing left to show is that $\text{dg } A \leq 0'$. Since $A(n) = (\delta_{n+1})_n$, it suffices to prove that $\text{dg } \{\delta_n\} \leq 0'$. We do this by showing that the degree of the construction is $\leq 0'$. This only makes sense, however, if the objects constructed at each step are finite objects. We therefore replace $T^s_i$ by an algorithm $I^s_i$ for $T^s_i$. We let $I^0_0$ be any algorithm for Id; and when we set $T^s_i$ equal to a previously defined $T^t_j$, we let $I^s_i = I^t_j$. The only remaining case is that $T^s_i$ is a splitting subtree of a previously defined tree. From the definition of splitting subtrees, we see that we can actually compute an algorithm for the I-splitting subtree of T for $\delta$ from I, $\delta$, and an algorithm for T. We take this algorithm to be $I^s_i$.

It is clear that oracles are needed in the construction only to obtain k and $\delta_{s+1}$ at step s+1. Now '$\delta$ has a proper extension on [J] incompatible with [I]' is an RE relation of $\delta$, J, I, since it can be written

$$\exists \alpha (\delta \subset \alpha \text{ \& } \delta \neq \alpha \text{ \& } \exists \gamma ([I](\gamma) = \alpha)$$
$$\text{\& } \exists i \exists n ([I](i) = n \text{ \& } i < \text{lh}(\alpha) \text{ \& } \alpha_i \neq n)).$$

An oracle for this RE relation suffices to obtain k. We can then find $\delta_{s+1}$ by means of the Selection Theorem. Q.E.D.

## 12. Simple Sets

The only RE degrees we have met so far are 0 and $0'$. This suggests <u>Post's Problem</u>: are there any other RE degrees?

Post hoped to give a positive solution of his problem by constructing a large non-recursive RE set A and using the largeness of A to prove that $\text{dg } A < 0'$. Although he did not solve the problem, he produced some interesting results in this manner. We shall look at one of his results.

The largest sets are the cofinite sets; but they are useless for our purposes because they are recursive. We will therefore require our large sets to be coinfinite.

A <u>simple</u> set in X is a coinfinite RE set A in X such that $A^c$ has no infinite RE subset. Such a set cannot be recursive; for then $A^c$ would be an infinite RE subset of $A^c$.

To use simple sets to solve Post's Problem, we would have to prove that there is a simple set and that every simple set has degree $< 0'$. Post showed that a simple set exists, but that not every simple set has degree $< 0'$. These results are included in the following theorem.

<u>Theorem 1</u> (Dekker). If a is RE and $a \neq 0$, then there is a simple set having degree a.

  <u>Proof</u>. Let A be an RE set in N having degree a. Since A is non-recursive, it is infinite; so it has a listing F by the Listing Theorem. Define an RE set B by

$$n \in B \leftrightarrow \exists m(m > n \ \& \ F(m) < F(n)).$$

Then $n \in B$ iff there is a $k < F(n)$ such that $k \in A$ and $k$ is distinct from $F(0), F(1), \ldots, F(n-1)$. It follows that $B \leq_R A$. If $k$ is given and we choose $n > k$ so that $F(n)$ is as small as possible, then $n \notin B$. Hence $B$ is coinfinite.

If $n \in B^c$, the members of $A$ other than $F(0), F(1), \ldots, F(n)$ are all $> F(n)$; so if $k < F(n)$, then $k \in A$ iff $k$ is one of $F(0), F(1), \ldots, F(n)$. Then given $k$ and an $n \in B^c$ such that $k < F(n)$, we can decide if $k \in A$. Such an $n$ certainly exists, since $B^c$ is infinite and $F$ is one-one. Given $k$ and an oracle for $B$, we can find such an $n$ by trying $0, 1, \ldots$ . Thus $A \leq_R B$; so dg $B = a$.

Suppose that $B^c$ has an infinite RE subset $C$. Then for each $k$ there is, as above, an $n \in C$ such that $k < F(n)$; and we can actually compute such an $n$ from $k$ by the Selection Theorem. Hence by the above, $A$ is recursive. This contradiction shows that $B$ is simple. Q.E.D.

*Remark*. The set $B$ has the following property: if $B^c$ has an infinite subset RE in $D$, then $B \leq_R D$. For the last paragraph in the proof shows that $A \leq_R D$; and $B \leq_R A$. A simple set $B$ with this property is said to be *strongly simple*.

Although we cannot prove that every simple set has degree $< 0'$, we can prove a weaker result.

*Theorem 2* (Post). A simple set is not complete RE.

*Proof*. Let $A$ be a complete RE set in $N$. Let $\{I_n\}$ be a listing of Alg(N,N). Since $n \in W_{I_n}$ is an RE relation of n, there is a recursive function F such that $n \in W_{I_n} \leftrightarrow F(n) \in A$ for all n. Let X = Sub(N), and choose a recursive G by the Parameter Theorem so that $n \in W_{G(x)} \leftrightarrow F(n) \in x$ for all n and x. Let H(x) = F(n), where n is chosen so that $I_n = G(x)$. Then H is recursive by Lemma 1 of §2. Moreover, we have $H(x) \in A \leftrightarrow H(x) \in x$; whence

(1) $\quad x \subset A^c \rightarrow H(x) \notin A \cup x$.

Define a recursive L inductively by

$$L(n) = H(\{L(0), L(1), \ldots, L(n-1)\}).$$

Using (1), we easily verify that $L(n) \in A^c$ and that L is one-one. Hence the range of L is an infinite RE subset of $A^c$; so A is not simple. Q.E.D.

## 13. The Priority Method

Post's Problem was solved independently by Friedberg and Muchnik, who developed for that purpose an important method called the <u>priority method</u>. We shall use this method to prove another result which has the solution of Post's Problem as a corollary.

<u>Theorem</u> (Friedberg). There is an RE degree a such that $a \neq 0$ and $a' = 0'$.

<u>Corollary</u>. There is an RE degree a such that $a \neq 0$ and $a \neq 0'$.

We will prove the theorem by constructing an RE set A in N and setting $a = dg\ A$. At step s, we will put a finite number of numbers into A. Then A will consist of all numbers which are put into A at some step. A difference from previous proofs is that our construction will be recursive; i.e., given s, we will be able to compute what is done at step s.

We let $A^s$ be the finite set of numbers which are put into A before step s. Since the construction is recursive, $n \in A^s$ is a recursive relation of n, s. Since $n \in A \leftrightarrow \exists s (n \in A^s)$, A is RE.

Since $a' \geq 0'$ always holds, we must insure that $a \neq 0$ and $a' \leq 0'$; equivalently, that A is not recursive and that $dg\ A^* \leq 0'$. This leads to the conditions:

$(1_I)$ $A \neq [I]$.

$(2_{I,k})$ Decide $k \in W_I^A$.

# THE PRIORITY METHOD

(As we shall see, $(2_{I,k})$ will have to be interpreted differently from the corresponding condition in §10.) Let $\{R_n\}$ be a listing of the space of conditions.

Our idea for insuring that $A \neq [I]$ is to pick a k and insure that $A(k) \neq [I](k)$. If $[I](k) = Fa$, we put k into A so that $A(k) = Tr$; if $[I](k) \neq Fa$, we do nothing about k, so that $A(k) = Fa$.

Since our construction is to be recursive, we must actually compute $[I](k)$. If $[I](k)$ is undefined, this computation will go on forever. This will not matter as far as this k is concerned; for if $[I](k)$ is undefined, we are to do nothing about k. The difficulty is that we will never get to the next condition.

Our solution of this difficulty is to keep coming back to each condition. Specifically, if s is in row n (in the terminology of §2), we devote step s to $R_n$. In particular, if $R_n$ is $A \neq [I]$, we do s steps in the computation of $[I](k)$ at step s. Thus if $[I](k)$ is defined, we will eventually compute its value.

Now suppose that we are trying to decide $k \in W_I^A$ at step s. We want to compute $W_I^A(k)$; but this is impossible, since A is not yet fully defined. However, we do have the finite approximation $A^s$ to A. We cannot even compute $W_I^{A^s}(k)$. However, we can do s steps in the computation of $[I]^{A^s}(k)$; and this may lead us to the conclusion that $k \in W_I^{A^s}$.

Even if we discover at step s that $k \in W_I^{A^s}$, it will do us no good unless we have $k \in W_I^A$. By the Use Principle, we will have $k \in W_I^A$ if $A(r) = A^s(r)$ for every r used in the computation of $[I]^{A^s}(k)$. Because of the way A is constructed, we can express this differently. We say that r is used negatively in the computation of $[I]^{A^s}(k)$ if r is used in that computation and $r \notin A^s$. Then we will have $k \in W_I^A$ if no r used negatively in the computation of $[I]^{A^s}(k)$ is put into A after step s.

If we discover that $k \in W_I^{A^s}$ at step s, we form the finite set x of numbers used negatively in the computation of $[I]^{A^s}(k)$. We call x an n-requirement, where n is the number such that $R_n$ is $(2_{I,k})$. The object of this requirement is to remind us that we do not wish to put the members of x into A.

Now suppose that we have created an n-requirement x, and that we later wish to put some $r \in x$ into A in order to satisfy a condition $R_m$ of type $(1_I)$. The solution to this dilemma is the crux of the priority method: we give priority to the lower numbered condition. Hence if $n < m$, we keep r out of A; if $m < n$, we put r into A.

If we violate the procedure for satisfying $R_n$ in order to follow the procedure for satisfying $R_m$, we say that $R_n$ is injured by $R_m$. When this happens, we can start again on the procedure for satisfying $R_n$. The reason that $R_n$ is eventually satisfied is that it can only be injured by a finite number of other requirements, viz., the $R_m$ for $m < n$.

If we create a requirement x at step s, we will have $x \cap A^s = \emptyset$. If $t > s$ and $x \cap A^t = \emptyset$, we say that x is <u>active</u> at step t; otherwise, x is <u>inactive</u> at step t. (The idea is that an inactive requirement has failed to serve its purpose and hence can be ignored.)

Now we explain how to select the argument k used in insuring $(1_I)$. If $R_n$ is $(1_I)$, we choose k in row n; this prevents different conditions from interfering with one another. In addition, we want to avoid having k kept out of A by requirements; so we choose k different from all numbers in active m-requirements with $m < n$.

We now describe step s. Let s be in row n, and first suppose that $R_n$ is $(1_I)$. We do nothing at this step unless: (a) no number in row n is in $A^s$; (b) there is a number $k < s$ such that k is in row n, k is in no active m-requirement with $m < n$, and $[I]_s(k) = Fa$. In this case, we choose the smallest such k and put k into A.

Now suppose that $R_n$ is $(2_{I,k})$. We do nothing unless there is no active n-requirement and $k \in W^{A^s}_{I,s}$. In this case, we create an n-requirement consisting of all numbers used negatively in the computation of $[I]^{A^s}(k)$.

This completes the description of the construction. We leave it to the reader to verify that it is indeed recursive.

If x is a requirement and $x \cap A = 0$ (so that x is active at every step after it is created), we say that x is _permanent_; otherwise, x is _temporary_.

If we put a member k of an active m-requirement x into A at step s, and k is in row n, then $n \leq m$. For each n, at most one number in row n is put into A. From these facts it follows that there are only finitely many temporary m-requirements. Since we never have two active m-requirements at the same time, there can be at most one permanent m-requirement. Hence there are only finitely many m-requirements.

Now we shall show that $R_n$ is satisfied. First suppose that $R_n$ is $(1_I)$. If a number k in row n is put into A, it is because we have discovered that $[I](k) = Fa$. Since $A(k) = Tr$, $(1_I)$ holds. Now suppose that no number in row n is in A. Pick k in row n so that k belongs to no m-requirement with $m < n$. Since $A(k) = Fa$, it will suffice to show that $[I](k) = Fa$ leads to a contradiction. Pick s in row n so large that $k \leq s$ and $[I]_s(k) = Fa$. Then some number in row n is put into A at step s, a contradiction.

Now suppose that $R_n$ is $(2_{I,k})$. The sense in which $R_n$ is satisfied is the following: we have $k \in W_I^A$ iff there is a permanent n-requirement. For suppose that a permanent n-requirement x is created at step s. Then $k \in W_I^{A^s}$. Since $x \cap A = 0$, no number used negatively in the computation of $[I]^{A^s}(k)$ is later put into A; so $k \in W_I^A$. Now let $k \in W_I^A$. Since $\lim A^s = A$, the Use Principle shows that $k \in W_{I,s}^{A^s}$ fo all sufficiently large s. Hence we can choose s in row n so large that $k \in W_{I,s}^{A^s}$ and every temporary n-requirement is inactive at step s. Either there is an active n-requirement at step s, or an n-requirement is created at step s. In either case, we have a permanent n-requirement.

We can now prove that $\text{dg } A^* \leq 0'$. Let $\langle I,k \rangle \in B_s$ if there is an active n-requirement at step s, where n is the number such that $R_n$ is $(2_{I,k})$. Since $\{B_s\}$ is clearly recursive, it suffices by the Limit Lemma to show that $\lim B_s = A^*$. If $\langle I,k \rangle \in A^*$, then there is a permanent n-requirement; so $\langle I,k \rangle \in B_s$ for all sufficiently large s. If $\langle I,k \rangle \notin A^*$, then all n-requirements are temporary; since there are only finitely many of them, $\langle I,k \rangle \notin B_s$ for all sufficiently large s. Q.E.D.

## 14. The Splitting Theorem

We shall now use the priority method to prove a result which gives some additional information about RE degrees.

<u>Splitting</u> <u>Theorem</u> (Sacks). Let C be an RE set and let D be a simple set. Then C is the union of disjoint RE sets A and B such that D is recursive in neither A nor B.

In order to translate this result into the language of degrees, we need a lemma.

<u>Lemma</u>. If A and B are disjoint RE sets, then $dg(A \cup B) = dg\,A \cup dg\,B$.

<u>Proof</u>. We must show that $A \cup B$ is recursive in A, B and that A and B are recursive in $A \cup B$. The former is evident. To show that A is recursive in $A \cup B$, it suffices by the reltivized Complementation Theorem to show that A and $A^c$ are RE in $A \cup B$. Since A is RE, it is RE in $A \cup B$. Now $A^c = (A \cup B)^c \cup B$; $(A \cup B)^c$ is recursive in $A \cup B$; and B is RE. It follows that $A^c$ is RE in $A \cup B$. A similar proof shows that B is recursive in $A \cup B$. Q.E.D.

We can now prove some corollaries of the Splitting Theorem.

<u>Corollary</u> <u>1</u>. Let c and d be RE degrees such that $d \neq 0$. Then there are RE degrees a and b such that $c = a \cup b$, $d \not\leq a$, and $d \not\leq b$.

<u>Proof</u>. Using Theorem 1 of §12, choose a simple set D with $dg\,D = d$. Let C be an RE set with $dg\,C = c$, and let

A and B be as in the theorem. Taking a = dg A and b = dg B, c = a ∪ b by the lemma; and clearly d ≰ a and d ≰ b.   Q.E.D.

Corollary 2. If c is a non-zero RE degree, then there are RE degrees a and b such that c = a ∪ b, 0 < a < c, 0 < b < c, and a|b.

Proof. Take d = c and let a and b be as in Corollary 1. From a ≤ a ∪ b = c and c ≰ a we conclude that a < c. Similarly, b < c. If a = 0, b = a ∪ b = c; so a ≠ 0. Similarly, b ≠ 0. If a and b are comparable, then a ∪ b = c is equal to either a or b; so a|b.   Q.E.D.

Corollary 3 (Friedberg-Muchnik). There are incomparable RE degrees.

Corollary 4 (Muchnik). No RE degree is minimal.

Corollary 5. If d is RE and 0 < d < 0', then there is an RE degree incomparable with d.

Proof. Let c = 0' and let a and b be as in Corollary 1. We claim that at least one of a and b is incomparable with d. If not, then a ≤ d and b ≤ d; so 0' = a ∪ b ≤ d, a contradiction.   Q.E.D.

Now we turn to the proof of the Splitting Theorem. We may suppose that C and D are sets in N. We shall first show how we can make use of these RE sets in a recursive construction. Define

$$k \in W_I^s \leftrightarrow k \leq s \ \& \ k \in W_{I,s}$$

for $I \in \mathrm{Alg}(N,N)$. For fixed I, the $W_I^s$ are finite sets which increase with s and have $W_I$ as their union. Moreover, $W_I^s$ can be computed from s and I. We now fix indices J of C and K of D, and set $C^s = W_J^s$, $D^s = W_K^s$. Then at step s, we may use $C^s$ and $D^s$ in our construction.

We insure that A and B are disjoint by never putting a number into both A and B; and we insure that A and B are subsets of C by only putting members of $C^s$ into A and B at step s. The rest is insured by the conditions:

$(1_k)$ $k \in C \rightarrow k \in A \cup B$;
$(2_I)$ $D \neq [I]^A$;
$(3_I)$ $D \neq [I]^B$.

Let $\{R_n\}$ be a listing of the space of conditions. If $R_n$ is $(2_I)$, we say n is an <u>A-number</u>; if $R_n$ is $(3_I)$, we say n is a <u>B-number</u>.

Our idea for satisfying $(2_I)$ is to try to make $[I]^A(k) = $ Fa for various k until we succeed for some k which is in D. We cannot make $[I]^A(k) = $ Fa for infinitely many k without obtaining a k in D; for the set of such k will be RE and D is simple. If we succeed in making $[I]^A(k) = $ Fa for only finitely many k, then some k for which we fail will be in the infinite set $D^c$; so we will have $[I]^A(k) \neq$ Fa $= D(k)$.

In the actual construction, we shall try to compute $[I]^{A^s}(k)$. If we find that $[I]^{A^s}(k) = $ Fa, we will create a requirement in an attempt to insure that $[I]^A(k) = Fa$. We will call this an n-<u>requirement</u> with <u>argument</u> k (where n is the number such that $R_n$ is $D \ne [I]^A$).

All of the above also holds with $(2_I)$ replaced by $(3_I)$ and A replaced by B.

Let x be an n-requirement created at step s, and suppose that n is an A-number. For $t > s$, we say that x is <u>active</u> at step t if $x \cap A^t = 0$; otherwise, x is <u>inactive</u> at step t. If $x \cap A = 0$, x is <u>permanent</u>; otherwise, x is <u>temporary</u>. Similar definitions hold with B in place of A.

We now describe step s. Let s be in row n, and first suppose that $R_n$ is $(1_k)$. We do nothing unless $k \in C^s$ and $k \notin A^s \cup B^s$. Suppose that this is the case. If k is in no active requirement, put k into A. Otherwise, choose n minimal such that k is in an active n-requirement. If n is an A-number, put k into B; if n is a B-number, put k into A. (Thus the condition $R_n$ with the smallest value of n is taking priority over the others.)

Now suppose that $R_n$ is $(2_I)$. Do nothing unless there is a number $k \le s$ such that: (i) no argument to an active n-requirement is $< k$ and in $D^s$; (ii) k is not the argument of an active n-requirement; (iii) $[I]_\epsilon^{A^s}(k) = Fa$. In this case, we pick the smallest such k and create an n-requirement with argument k consisting of all numbers used negatively in the computation of

$[I]^{A^s}(k)$. If $R_n$ is $(3_I)$, we proceed similarly with A replaced by B.

We shall prove by induction on n that there are only finitely many n-requirements. If n is an A-number, and a number k in an active n-requirement is put into A, then k must also belong to an active m-requirement with $m < n$. The same is true with A replaced by B. Using this and the induction hypothesis, we see that there are only finitely many temporary n-requirements.

Let $E_n$ be the set of arguments of permanent n-requirements. Let $E_n^s$ be the set of arguments of n-requirements active at step s. Then $k \in E_n^s$ is a recursive relation of k, s, n. By the above, we can choose $s_0$ so that every temporary n-requirement is inactive at step $s_0$. Then

$$k \in E_n \leftrightarrow \exists s(s \geq s_0 \ \& \ k \in E_n^s).$$

It follows that $E_n$ is RE.

Suppose that $E_n$ is infinite. Since D is simple, there is a $k \in D \cap E_n$. If s is sufficiently large, then at step s there is an active n-requirement with argument k and $k \in D^s$. Then no n-requirement with an argument $> k$ can be created at step s. We conclude that $E_n$ must be finite. Now there can never be two active n-requirements with the same argument at the same time; so there cannot be two permanent n-requirements with the same argument. Thus there are only finitely many permanent n-requirements, and hence only finitely many n-requirements.

Now we can show that $R_n$ is satisfied. Suppose that $R_n$ is $(1_k)$. Suppose that $k \in C$, and choose s in row n so large that $k \in C^s$. Then either $k \in A^s \cup B^s$, or k is put into A or B at step s. In any case, $k \in A \cup B$.

Now let $R_n$ be $(2_I)$. We suppose that $[I]^A = D$ and derive a contradiction. If an n-requirement with argument k is created at step s, then $[I]^{A^s}_s(k) = Fa$; and if the requirement is permanent, then $[I]^A(k) = Fa$ and hence $k \notin D$. Thus $E_n \subset D^c$. Since $E_n$ is finite and D is simple, we can choose $k \notin D \cup E_n$. Then $[I]^A(k) = D(k) = Fa$; so $[I]^{A^s}_s(k) = Fa$ for sufficiently large s. Choose s in row n so large that all n-requirements are created before step s; all temporary n-requirements are inactive at step s; $k \leq s$; and $[I]^{A^s}_s(k) = Fa$. If an n-requirement is active at step s, it must be permanent. Thus if it has argument k', then $k' \in E_n$; so $k' \in D^c$; so $k' \notin D^s$. Moreover, the requirement cannot have argument k, since $k \notin E_n$. It follows that an n-requirement is created at step s; and this contradicts the choice of s.

If $R_n$ is $(3_I)$, we proceed similarly with A replaced by B. Q. E. D.

## 15. Maximal Sets

We will show that Post's method cannot be used to solve his problem if <u>large</u> is taken to mean <u>having</u> <u>few</u> <u>RE</u> <u>supersets</u>. To prove this, we take the strongest possible notion of largeness and show that we still cannot prove that a large set has degree $<$ 0'.

If A is a coinfinite RE set, then among the RE supersets of A are the sets obtained from A by adding a finite number of elements and the cofinite sets which include A. We propose to exclude all others.

A <u>maximal</u> set is a coinfinite RE set A such that for every RE set B including A, either B - A or $B^c$ is finite.

A slight reformulation of the definition is useful. As B runs through all RE sets, A ∪ B runs through all RE sets including A. Moreover, (A ∪ B) - A = B ∩ $A^c$ and (A ∪ B)$^c$ = $B^c$ ∩ $A^c$. Thus a coinfinite RE set A is maximal iff for every RE set B, either B ∩ $A^c$ or $B^c$ ∩ $A^c$ is finite.

A maximal set is simple and hence non-recursive. For suppose that $A^c$ has an infinite RE subset C. Let F be a listing of C, and set G(n) = F(2n). Then the range B of G is an RE set such that both B ∩ $A^c$ and $B^c$ ∩ $A^c$ are infinite.

The existence of a maximal set was first proved by Friedberg. The result which we want is that there is a maximal set of degree 0'; this was first proved by Yates. We shall prove a still stronger result.

72

# MAXIMAL SETS

**Theorem** (Martin). An RE degree a is the degree of a maximal set iff a' = 0''.

If a is RE, then $a \leq 0'$ and hence $a' \leq 0''$; so a' = 0'' is equivalent to $a' \geq 0''$. We begin by characterizing the degrees a such that $a' \geq 0''$.

If F and G are functions from N to N, we say F <u>dominates</u> G if $F(n) > G(n)$ for all sufficiently large n. A function from N to N is <u>dominant</u> if it dominates every recursive function from N to N.

**Lemma 1.** For any degree a, $a' \geq 0''$ iff there is a dominant function of degree $\leq a$.

*Proof.* Suppose that F is dominant and dg F $\leq$ a. Let A be a complete $\pi_2$ set in N. Then

$$n \in A \leftrightarrow \forall k \, \exists r (\langle n,k,r \rangle \in B)$$

with B recursive. Set

$$n \in A_s \leftrightarrow (\forall k \leq s)(\exists r \leq F(s))(\langle n,k,r \rangle \in B).$$

Then dg $\{A_s\}$ $\leq$ dg F $\leq$ a. We show that A = lim $A_s$; it follows by the Limit Lemma that 0'' = dg A $\leq$ (dg $\{A_s\}$)' $\leq$ a'.

If $n \notin A$, then there is a k such that $\langle n,k,r \rangle \notin B$ for all r; so $n \notin A_s$ for $s \geq k$. If $n \in A$, we can define a recursive function G by

$$G(s) = \max_{k \leq s} \mu r (\langle n,k,r \rangle \in B).$$

For large s, G(s) < F(s) and hence $n \in A_s$. Thus in either case, $\lim A_s(n) = A(n)$.

Now let $a' \geq 0''$. Then $dg(Tot) = 0'' \leq a'$. Hence by the Limit Lemma, there is a sequence $\{A_n\}$ such that $dg\{A_n\} \leq a$ and $\lim A_n = Tot$. For each n and I, there is an s such that

(1)    $[I](n) = s \lor (s \geq n \; \& \; I \notin A_s)$.

For if $I \in Tot$, we can take $s = [I](n)$; while if $I \notin Tot$, then $I \notin A_s$ for all large s. Now (1) is RE in $\{A_n\}$; so by the relativized Selection Theorem, there is a G of degree $\leq a$ such that for all n and I, $G(n,I)$ is an s for which (1) holds.

Let $\{I_n\}$ be a listing of $Alg(N,N)$, and set

$$F(n) = \max_{i \leq n} G(n, I_i) + 1.$$

Then $dg\, F \leq a$. Let H be a recursive function from N to N. Then $H = [I_i]$ with $I_i \in Tot$. For large n, $I_i \in A_s$ for every $s \geq n$; so $G(n, I_i) = [I_i](n) = H(n)$. Hence for large $n \geq i$, $H(n) < F(n)$. Q.E.D.

If A is a coinfinite set in N, $L_A$ is the function from N to N such that $L_A(0), L_A(1), \ldots$ are the members of $A^c$ in increasing order. Then $dg\, L_A = dg\, A$. For given an oracle for A, we can compute $L_A(0), L_A(1), \ldots$ in turn. Conversely, given an oracle for $L_A$, we can decide if $k \in A$ by computing $L_A(0), L_A(1), \ldots$ until we come to k or to a number larger than k.

Lemma 2. If A is a maximal set in N, then $L_A$ is dominant.

## MAXIMAL SETS

Proof. Let F be a recursive function from N to N. Define a recursive H by

$$H(k) = \mu n(F(2n) + F(2n+1) + n \geq k),$$

and define an RE B by

$$k \in B \leftrightarrow (\forall r < k)(H(r) = H(k) \rightarrow r \in A).$$

If $H(k) = n$ for some $k \in A^c$, then exactly one such k (viz., the smallest one) is in B. Since either $B \cap A^c$ or $B^c \cap A^c$ is finite, it follows that for large n, there is at most one $k \in A^c$ for which $H(k) = n$. This implies that there is an r such that for all n, there are at most $n + r$ numbers $k \in A^c$ such that $H(k) \leq n$. These include all numbers $k \in A^c$ such that $k \leq F(n) + F(2n+1) + n$. Hence

$$F(2n) + F(2n+1) + n \leq L_A(n+r).$$

If $n > r$, $L_A(n+r) < L_A(2n) < L_A(2n+1)$. It follows that $L_A$ dominates F. Q.E.D.

We can now prove half of the theorem. Suppose that a is the degree of a maximal set A. We may suppose that A is a set in N. Then $a' = (dg\ A)' = (dg\ L_A)' \geq 0''$ by Lemmas 1 and 2. Since a is RE, $a' = 0''$.

For the converse, we will have to produce an RE set of a specified degree. We develop a method for doing this.

A <u>restricting</u> function is a recursive function F from N to N such that F(s) goes to infinity with s. For such an F we define

$$\nu_F(n) = \mu s(\forall t \geq s)(F(t) > n).$$
If $\nu_F(n) \leq G(n)$ for all $n$, then $\nu_F \leq_R G$; for
$$\nu_F(n) = \mu s(\forall t \leq G(n))(t \geq s \to F(t) > n).$$

The construction of an RE set $A$ in $N$ is F-<u>restricted</u> if whenever $n$ is put into $A$ at step $s$, then $n \geq F(s)$. If $F$ is restricting and the construction of $A$ is F-restricted, then $A \leq_R \nu_F$. For if $n$ is put into $A$ at step $s$, then $n \geq F(s)$; so $s < \nu_F(n)$. Thus $n \in A \leftrightarrow n \in A^{\nu_F(n)}$.

If we are constructing an RE set $A$ in $N$, we write $L_s$ for $L_{A^s}$. The construction of $A$ is F-<u>supported</u> if for each $s$, some number $\leq L_s(F(s))$ is put into $A$ at step $s$.

If $F$ is restricting and the construction of $A$ is F-supported, then $\nu_F \leq_R A$. To see this, let $G(n)$ be the smallest $s$ such that every number $\leq L_A(n)$ which is in $A$ is in $A^s$. Then $G \leq_R A$; so it suffices to show that $\nu_F \leq_R G$. For this it is enough to show that $\nu_F(n) \leq G(n)$; i.e., that if $s \geq G(n)$, then $F(s) > n$. At step $s$, we put into $A$ a number $k \leq L_s(F(s))$. Since $s \geq G(n)$, $k \notin A^{G(n)}$; so $k > L_A(n)$. Since $A^s \subset A$, $L_s(F(s)) \leq L_A(F(s))$. Combining these three inequalities, $L_A(F(s)) > L_A(n)$; so $F(s) > n$.

<u>Example</u>. Let $F$ be a listing of an RE set $A$ in $N$. Since $F$ is one-one, it is restricting. We can construct $A$ by putting $F(s)$ into $A$ at step $s$. Since $F(s) \leq L_s(F(s))$, this construction is F-restricted and F-supported; so dg $A$ = dg $\nu_F$.

Lemma 3. If a is RE and $a' = 0''$, then there is a restricting function F such that $\nu_F$ is dominant and dg $\nu_F = a$.

Proof. Let A be an RE set in N of degree a. Since A is non-recursive, it is infinite; so it has a listing G by the Listing Theorem. By Lemma 1, there is a dominant function H recursive in A. By the Modulus Lemma, there is a recursive sequence $\{H_n\}$ with the limit H and a modulus M of $\{H_n\}$ which is recursive in A.

Let $F(s)$ be the smallest number $n \leq G(s)$ such that $H_s(n) \neq H_{s+1}(n)$, if there is such an n; otherwise, let $F(s) = G(s)$. Clearly F is recursive. If $s \geq M(m)$ for all $m \leq n$ and $s \geq \nu_G(n)$, then $F(s) > n$. It follows that F is restricting and that

$$\nu_F(n) \leq \max_{m \leq n} M(m) + \nu_G(n).$$

Hence $\nu_F$ is recursive in M, $\nu_G$ and therefore in A. Since $F(s) \leq G(s)$, $\nu_G(n) \leq \nu_F(n)$; so $A \leq_R \nu_G \leq_R \nu_F$. Thus dg $\nu_F = a$.

It remains to show that $\nu_F$ is dominant. If L is recursive, then $H_{L(n)}(n) < H(n)$ for all sufficiently large n. For any such n, there is an $s \geq L(n)$ such that $H_s(n) \neq H_{s+1}(n)$ and hence $F(s) \leq n$. This implies that $L(n) \leq \nu_F(n)$. Q.E.D.

We now reformulate the definition of a maximal set. If A is a set in the domain of F, and if there is a finite subset $A_0$ of A such that F is constant on $A - A_0$, then we say that

F is <u>almost</u> <u>constant</u> on A.  Then a coinfinite RE set A is maximal iff for every RE B, the function B is almost constant on $A^c$.

An n-<u>string</u> is a string of length n.  We fix a listing $\{I_n\}$ of $Alg(N,N)$ and let $S_n(k)$ by the n-string

$$W_{I_0}(k), W_{I_1}(k), \ldots, W_{I_{n-1}}(k).$$

Then a coinfinite RE set A in N is maximal iff each $S_n$ is almost constant on $A^c$.

We order the n-strings by setting $\alpha < \beta$ if $\alpha_i < \beta_i$, where i is the smallest number such that $\alpha_i \neq \beta_i$.  It is easy to check that this is a linear ordering of the n-strings. Moreover

$$(2) \qquad m \leq n \ \& \ S_m(i) < S_m(j) \rightarrow S_n(i) < S_n(j),$$

so that

$$(3) \qquad m \leq n \ \& \ S_n(j) \leq S_n(i) \rightarrow S_m(j) \leq S_m(i).$$

Our idea for constructing a maximal set A is to put i into A whenever we find a $j > i$ such that $S_n(j) > S_n(i)$.  If $\alpha$ is the largest n-string such that $S_n(j) = \alpha$ for infinitely many j, then every i with $S_n(i) < \alpha$ will be put into A.  This implies that $S_n$ is almost constant on $A^c$.

Two modifications are necessary.  In order to be sure that A is coinfinite, we will only put i into A because $S_n(j) > S_n(i)$ if there are at least n numbers less than i which have not been put into A.

The second modification is needed to make the construction recursive. We let $S_n^s(k)$ be the n-string

$$W_{I_0}^s(k), W_{I_1}^s(k), \ldots, W_{I_{n-1}}^s(k).$$

Then (2) and (3) still hold if we insert superscripts s on S everywhere. Moreover, $S_n^s(k)$ is a recursive function of s, n, k. Since Fa < Tr,

(4) $\quad s \leq t \rightarrow S_n^s(k) \leq S_n^t(k);$

and clearly

(8) $\quad \lim S_n^s(k) = S_n(k).$

We then modify the above procedure by replacing $S_n$ by $S_n^s$ at step s of the construction.

We now complete the proof of the theorem. Let a be an RE degree such that a' = 0''; we shall construct a maximal set A in N of degree a. Let F be as in Lemma 3, and let F'(x) = F(x) + 1. Our construction will be F-restricted and F'-supported, so that $\nu_{F'} \leq_R A \leq_R \nu_F$. But $\nu_F(n) = \nu_{F'}(n+1)$; so $\nu_F \leq_R \nu_{F'}$; so dg A = dg $\nu_F$ = a. Of course making the construction F-restricted and F'-supported will require further modification of our procedure.

At step s, we say that j n-<u>holds</u> i if $L_s(n) \leq i < j$, $S_n^s(i) < S_n^s(j)$, and i, j $\notin A^s$. This implies that $j \leq s$; for if j > s, $S_n^s(j)$ = (Fa, ..., Fa) is the smallest n-string. If some j n-holds i, we say that i is n-<u>held</u>. If i is n-held for some n, we say that i is <u>held</u>.

At step s, we put into A all numbers which are held and are $\geq F(s)$. If none of these numbers is $\leq L_s(F'(s))$, then we also put $L_s(F'(s))$ into A. Since $L_s(F'(s)) \geq F'(s) > F(s)$, the construction is F-restricted and F'-supported.

If i is held at step s, the <u>holder</u> of i is the smallest number $j > i$ such that $j \notin A^s$ and j is not held. Then this j is not put into A at step s. We need only verify this if $j = L_s(F'(s))$. Then, by the definition of j, $L_s(F(s))$ is held. Since $L_s(F(s)) \geq F(s)$, $L_s(F(s))$ is put into A; so j is not.

We show that if i is n-held at step s and j is its holder, then j n-holds i. Clearly $L_s(n) \leq i < j$. Let k be the largest number which n-holds i. If $r > k$ and $r \notin A^s$, then $S_n^s(k) > S_n^s(r)$; so $S_m^s(k) \geq S_m^s(r)$ for all m by (2) and (3). This shows that k is not held; so $j \leq k$. But k cannot n-hold j; so $S_n^s(j) \geq S_n^s(k) > S_n^s(i)$. Thus j n-holds i.

We now show by induction on n that $\lim L_s(n) = L(n)$ exists. Using the induction hypothesis, choose $s_0 \geq \nu_F(n)$ so that for $s \geq s_0$, $L_s(m) = L(m)$ for all $m < n$. We show first that

(6) $\quad s \geq s_0 \ \& \ L_{s+1}(n) \neq L_s(n) \rightarrow S_n^{s+1}(L_{s+1}(n)) > S_n^s(L_s(n))$.

Since $L_s(m) = L_{s+1}(m)$ for $m < n$ and $L_s(n) \neq L_{s+1}(n)$, $L_s(n)$ is the smallest number put into A at step s. Since $s \geq \nu_F(n)$, $F'(s) > F(s) > n$; so $L_s(n)$ is put into A because it is held and is $\geq F(s)$. Thus all numbers $\geq L_s(n)$ which are held are put into A. Hence the smallest number $\geq L_s(n)$ which is not in

$A^s$ and is not put into A at step s is the holder of $L_s(n)$. Thus $L_{s+1}(n)$ is the holder of $L_s(n)$. By (2), $L_s(n)$ is n-held; so by the above, $L_{s+1}(n)$ n-holds $L_s(n)$. Then by (4),

$$S_n^{s+1}(L_{s+1}(n)) \geq S_n^s(L_{s+1}(n)) > S_n^s(L_s(n)),$$

proving (6).

If $L_{s+1}(n) = L_s(n)$, then $S_n^{s+1}(L_{s+1}(n)) \geq S_n^s(L_s(n))$ by (4). From this and (6), we see that for $s \geq s_0$, $S_n^s(L_s(n))$ increases with s. Since there are only finitely many n-strings, it eventually becomes constant. From this and (6), we see that $L_s(n)$ eventually becomes constant; so $\lim L_s(n)$ exists.

For all sufficiently large s, $L(n) = L_s(n) \notin A^s$; so $L(n) \notin A$. If $m \neq n$, $L_s(m) \neq L_s(n)$ for all s; so $L(m) \neq L(n)$. These facts show that A is coinfinite.

It remains to show that $S_n$ is almost constant on $A^c$. Let $\alpha$ be the smallest n-string such that $S_n(i) = \alpha$ for infinitely many $i \in A^c$. It will suffice to assume that $S_n(i) > \alpha$ for infinitely many $i \in A^c$ and derive a contradiction.

Let B be the set of $i \in A^c$ such that $S_n(i) = \alpha$ and $L_A(n) \leq i$. Then B is infinite. Let $C^s$ be the set of i such that $S_n^s(i) = \alpha$ and i is n-held at step s. We show that

(8)     $s \leq t \rightarrow B \cap C^s \subset C^t$.

It will suffice to show that $B \cap C^s \subset C^{s+1}$. Let $i \in B \cap C^s$, and let j be the holder of i at step s. Since $i \in A^c$ and j is a holder at step s, $i, j \notin A^{s+1}$. We have $L_{s+1}(n) \leq L_A(n) \leq$

$i < j$.  Since $S_n^s(i) = S_n(i) = \alpha$, $S_n^{s+1}(i) = \alpha$ by (4) and (5). Now $j$ n-holds $i$ at step $s$; so $S_n^{s+1}(i) = S_n^s(i) < S_n^s(j) \leq S_n^{s+1}(j)$ by (4).  Thus $i \in C^{s+1}$.

Define $D$ by $\langle i,s \rangle \in D \leftrightarrow i \notin B \vee i \in C^s$.  We claim that $D$ is RE.  Since $i \in C^s$ is a recursive relation of $i,s$, it suffices to verify that $B^c$ is RE.  Now $B^c$ is the union of the four sets $A$, $[i: S_n(i) > \alpha]$, $[i: i \in A^c \ \& \ S_n(i) < \alpha]$, and $[i: i < L_A(n)]$.  The first two are RE, since

$$S_n(i) > \alpha \leftrightarrow \exists s(S_n^s(i) > \alpha)$$

by (4) and (5); and the last two are finite.  Thus $B^c$ is RE.

For each $i$ there is an $s$ such that $\langle i,s \rangle \in D$.  This is clear if $i \notin B$; so suppose that $i \in B$.  Since there are infinitely many $j \in A^c$ such that $S_n(j) > \alpha$, we can choose such a $j$ with $j > i$.  Then $S_n(i) < S_n(j)$; so by (5), $S_n^s(i) < S_n^s(j)$ for all sufficiently large $s$.  For all $s$, we have $i, j \notin A^s$ and $L_s(n) \leq L_A(n) \leq i < j$.  Thus for large $s$ we have $i \in C^s$ and hence $\langle i,s \rangle \in D$.

By the Selection Theorem, there is a recursive $H$ such that $\langle i, H(i) \rangle \in D$ for all $i$.  Since $\nu_F$ is dominant, $H(i) < \nu_F(i)$ for all sufficiently large $i$.  Since $B$ is infinite, there is an $i \in B$ with $H(i) < \nu_F(i)$.  Let $s = H(i)$.  Since $i \in B$ and $\langle i,s \rangle \in D$, $i \in C^s$.  Since $s < \nu_F(i)$, there is a $t \geq s$ such that $F(t) \leq i$.  By (8), $i \in C^t$.  Thus $i$ is held at step $t$ and $i \geq F(t)$; so $i$ is put into $A$ at step $t$.  This contradicts $i \in A^c$.  Q.E.D.

## 16. Infinite Injury

We might try to improve the results of the last section by showing that no notion of large for RE sets can lead to a solution of Post's Problem. We have seen that we want our large sets to be coinfinite; and we certainly want any coinfinite RE set which includes a large set to be large. We would thus achieve our object if we could prove: every coinfinite RE set is included in a coinfinite RE set having degree $0'$.

We will show, however, that this result is false. If $A$ is a maximal set, then every coinfinite RE set including $A$ has the same degree as $A$. Thus we need only show that there is a maximal set not of degree $0'$. This follows from the last section and the following result.

Theorem (Sacks). There is an RE degree $a$ such that $a' = 0''$ and $a \neq 0'$.

If $A$ is a set in $X \times Y$, $A^{(x)}$ is the set of $y$ such that $\langle x, y \rangle \in A$. We say that $A$ is <u>piecewise recursive</u> if each $A^{(x)}$ is recursive. A <u>thick</u> subset of $A$ is a subset $B$ such that $A^{(x)} - B^{(x)}$ is finite for every $x$.

Lemma (Shoenfield). If $C$ is a piecewise recursive RE set in $X \times Y$, then there is a thick RE subset $A$ of $C$ such that $\text{dg } A \neq 0'$.

We first show that the lemma implies the theorem. Define

an RE C by $\langle I,k\rangle \in C \leftrightarrow (\forall r < k)(r \in W_I)$. Then

(1) $\quad\quad I \in \text{Tot} \rightarrow C^{(I)} = N,$
$\quad\quad\quad I \notin \text{Tot} \rightarrow C^{(I)}$ is finite.

Thus C is piecewise recursive. Let A be as in the Lemma and let $a = \text{dg } A$. Then $a \neq 0'$. Set $A_k(I) = A(I,k)$. By (1) and the thickness of A, we have $\lim A_k = \text{Tot}$. Hence by the Limit Lemma, $0'' = \text{dg}(\text{Tot}) \leq (\text{dg } A)' = a'$. Since a is RE, it follows that $a' = 0''$. Q.E.D.

Now we turn to the proof of the lemma. We take $X = Y = N$. Let D be a simple set in N. We insure that $\text{dg } A \neq 0'$ by insuring that D is not recursive in A.

For $I \in \text{Alg}(N \times N, N)$ we define

$$\langle k,r\rangle \in W_I^s \leftrightarrow \langle k,r\rangle \in W_{I,s} \ \& \ k \leq s \ \& \ r \leq s.$$

We then choose indices J and K of C and D respectively and set $C^s = W_J^s$, $D^s = W_K^s$. To insure that $A \subset C$, we only put members of $C^s$ into A at step s. In addition, we must insure that $C^{(m)} - A^{(m)}$ is finite for each m and that $[I]^A \neq D$ for all I. To set up the priorities, fix a listing $\{I_n\}$ of $\text{Alg}(N,N)$. Then making $C^{(m)} - A^{(m)}$ finite takes priority over making $[I_n]^A \neq D$ if $m < n$, while the reverse priority holds if $n \leq m$.

If $C^{(m)}$ is infinite, we will have to put infinite many pairs into A in order to make $C^{(m)} - A^{(m)}$ finite. In doing this, we may injure a condition $[I_n]^A \neq D$ infinitely often. For this reason, our construction is called an <u>infinite injury</u> construction.

# INFINITE INJURY

Our method of insuring that $[I_n]^A \neq D$ is similar to that used in the proof of the Splitting Theorem. Thus having computed that $[I_n]^{A^s}(k) = Fa$ for a suitable k, we create an n-requirement with argument k to insure that $[I_n]^A(k) = Fa$ unless the requirement is temporary. Because of the infinite injury situation, we will not be able to prove that there are only finitely many n-requirements. However, by using the piecewise recursiveness of C, we will be able to prove as before that there are only finitely many permanent n-requirements.

There is still the difficulty that a pair $\langle m,r \rangle$ in C may not get into A because at every step after it appears in C, it is in an active but temporary requirement. To prevent this, we do not allow $\langle m,r \rangle$ to be put in a requirement at step s if $\langle m,r \rangle \in C^s$, unless $\langle m,r \rangle$ is put into A at step s. Instead, we put into the requirement certain pairs not in $C^s$, at least one of which will have to be put into A before $\langle m,r \rangle$ is put into A.

We let $P_s(m,r)$ be the set of $\langle m',r' \rangle$ such that $m' < m$ and $\langle m',r' \rangle$ belongs to a requirement which is active at step s and contains $\langle m,r \rangle$. If x is a finite set in $N \times N$, $Q_s(x)$ is the smallest set such that $x \subset Q_s(x)$ and

(2) $\quad \langle m,r \rangle \in Q_s(x) \cap C^s \rightarrow P_s(m,r) \subset Q_s(x)$.

Clearly $Q_s(x)$ is included in the union of x and the requirements which are active at step s. Hence $Q_s(x)$ is finite and can be computed at step s; and if $x \cap A^s = \emptyset$, then $Q_s(x) \cap A^s = \emptyset$.

We now describe step s. It consists of two parts. In the first part, we put into A all $\langle m,r \rangle \in C^s - A^s$ which belong to no active n-requirement with $n \leq m$.

For the second part, let s be in row n. We do nothing unless there is a $k \leq s$ such that: (a) no argument to an active n-requirement is $< k$ and in $D^s$; (b) k is not the argument of an active n-requirement; (c) $[I_n]_s^{A^s}(k) = Fa$. In this case, we pick the smallest such k and let x be the set of pairs used negatively in the computation of $[I_n]^{A^s}(k)$. Let y be the set of $\langle m,r \rangle \in Q_s(x)$ such that either $\langle m,r \rangle \notin C^s$ or $\langle m,r \rangle$ has been put into A in the first part of this step. We then make y into an n-requirement with argument k.

We prove a series of facts leading to the conclusion that all of the conditions are satisfied.

(A) If an n-requirement y is created at step s, then y is permanent iff $y \cap A^{s+1} = 0$ and no $\langle m,r \rangle \in y$ with $m < n$ is in A.

This follows from the fact that if $y \cap A^{s+1} = 0$, then the first element of y (if any) to be put into A must be a pair $\langle m,r \rangle$ with $m < n$.

(B) If $\langle m,r \rangle \in C$, then $\langle m,r \rangle$ belongs to only finitely many requirements.

For if $\langle m,r \rangle \in C^s$, then $\langle m,r \rangle$ is not put into a requirement after step s except, possibly, at the step at which it is put into A.

# INFINITE INJURY

(C) If $\langle m,r \rangle \in C$, then $\langle m,r \rangle \in A$ iff $\langle m,r \rangle$ belongs to no permanent n-requirement with $n \leq m$.

A member of A can belong to no permanent requirement. Suppose $\langle m,r \rangle \notin A$. By (B), we can choose s so large that $\langle m,r \rangle \in C^s$ and every temporary requirement containing $\langle m,r \rangle$ is inactive at step s. Since $\langle m,r \rangle \in C^s - A^s$ and $\langle m,r \rangle$ is not put into A at step s, there is an n-requirement with $n \leq m$ which is active at step s and contains $\langle m,r \rangle$. This requirement must be permanent.

We let $E_n$ be the set of arguments of permanent n-requirements.

(D) $E_n$ and $C^{(n)} - A^{(n)}$ are finite.

We prove this by induction on n. Our first step is to show that $E_n$ is RE. Let $E_n^s$ be the set of arguments of permanent n-requirements created at step s. Since $k \in E_n \leftrightarrow \exists s(k \in E_n^s)$, it suffices to show that $k \in E_n^s$ is a recursive relation of k, s. Given k and s, we can decide if an n-requirement with argument k is created at step s; and, if so, we can find this requirement y. We must now decide whether y is permanent or temporary. By (A), it will suffice to decide for each $\langle m,r \rangle \in y$ with $m < n$ whether or not $r \in A^{(m)}$. By induction hypothesis, $C^{(m)} - A^{(m)}$ is finite. Since C is piecewise recursive, this implies that $A^{(m)}$ is recursive. Thus we can decide whether or not $r \in A^{(m)}$.

Now suppose that $E_n$ is infinite. Since D is simple, there is a $k \in D \cap E_n$. For large s, $k \in D^s$ and there is an n-requirement with argument k active at step s. For such s, no n-requirement with an argument $> k$ is created at step s. But this implies that $E_n$ is finite.

From the result just proved and the induction hypothesis, we conclude that $E_m$ is finite for $m \leq n$. Since two m-requirements with the same argument cannot be active at the same step, there cannot be two permanent m-requirements with the same argument. Hence there are only finitely many permanent m-requirements with $m \leq n$. It follows from (C) that $C^{(n)} - A^{(n)}$ is finite.

(E) If there is a permanent n-requirement with argument k, then $[I_n]^A(k) = Fa$.

We suppose that the requirement y is created at step s, and let x be as in the description of step s. It will clearly suffice to show that $x \cap A = \emptyset$. Suppose that $x \cap A \neq \emptyset$, so that $Q_s(x) \cap A \neq \emptyset$. Choose $\langle m,r \rangle \in Q_s(x) \cap A$ with m minimal. Since $x \cap A^s = \emptyset$, $Q_s(x) \cap A^s = \emptyset$; so $\langle m,r \rangle \notin A^s$. Since y is permanent, $\langle m,r \rangle \notin y$; so $\langle m,r \rangle \in C^s$ and $\langle m,r \rangle$ is not put into A at step s. It follows that at step s, $\langle m,r \rangle$ belongs to an active n'-requirement z with $n' \leq m$. Since $\langle m,r \rangle \in z \cap A$, z is temporary. Since z is active at some step, it follows from (A) that there is a $\langle m',r' \rangle \in z \cap A$ with $m' < n' \leq m$. Then $\langle m',r' \rangle \in P_s(m,r)$; so by (2), $\langle m',r' \rangle \in Q_s(x)$. This contradicts the choice of $\langle m,r \rangle$.

We let $P(m,r)$ be the union of the $P_s(m,r)$ for $s = 0, 1, \ldots$ . It follows from (B) that $P(m,r)$ is finite if $\langle m,r \rangle \in C$. We say a finite set $y$ in $N \times N$ is __closed__ if

$$\langle m,r \rangle \in y \cap C \rightarrow P(m,r) \subset y.$$

Then if $y$ is closed and $x \subset y$, then $Q_s(x) \subset y$ for all $s$.

(F) Every finite subset of $N \times N$ is included in a closed set.

Since the union of a finite number of closed sets is closed, it will suffice to show that every $\langle m,r \rangle$ belongs to a closed set. We do this by induction on $m$. If $\langle m,r \rangle \notin C$, then $\{\langle m,r \rangle\}$ is closed. Let $\langle m,r \rangle \in C$. For each $\langle m',r' \rangle \in P(m,r)$, choose a closed set containing $\langle m',r' \rangle$ by the induction hypothesis. The union of these sets and $\{\langle m,r \rangle\}$ is closed.

(G) If $[I_n]^A(k)$ is defined, then there **are** only finitely many n-requirements with argument $k$.

For large $s$, then computation of $[I_n]^{A^s}(k)$ is the same as the computation of $[I_n]^A(k)$. Hence there is a finite set $x$ such that for every $s$ for which $[I_n]^{A^s}(k)$ is defined, each pair used negatively in the computation of $[I_n]^{A^s}(k)$ is in $x$. By (F), there is a closed set $y$ including $x$. Every n-requirement with argument $k$ is included in $y$. Each time that such a requirement becomes inactive, some member of $y$ is put into $A$. Thus there are only finitely many temporary n-requirements with argument k; and there is at most one permanent n-requirement with argument k.

(H) $[I_n]^A \neq D$.

We assume that $[I_n]^A = D$ and derive a contradiction. Since $E_n$ is finite by (D) and D is simple, we can choose a $k \in (D \cup E_n)^c$. Then $[I_n]^A(k) = D(k) = Fa$. By (D) and (G), we can choose s in row n so large that: (a) every permanent n-requirement is active at step s; (b) every temporary n-requirement with an argument $\leq k$ is inactive at step s; (c) $[I_n]_s^{A^s}(k) = Fa$; (d) $k \leq s$. If at step s there is an active n-requirement with an argument $m \leq k$, then it is permanent; so $D(m) = [I_n]^A(m) = Fa$ by (E); so $m \notin D^s$. Also $m \neq k$, since $k \notin E_n$. These facts show that an n-requirement with an argument $\leq k$ is created at step s. But this is impossible by (a) and (b).

It follows from (D) and (H) that all of the conditions are satisfied.    Q.E.D.

We shall need some further facts about the construction just made in the next section. First, we show that A is recursive in C. We suppose that an oracle for C is given, and show how to compute whether or not $\langle m,r \rangle \in A$ by induction on m. If $\langle m,r \rangle \notin C$, then $\langle m,r \rangle \notin A$; so we suppose that $\langle m,r \rangle \in C$. We find an s such that $\langle m,r \rangle \in C^s$. A requirement created after step s can contain $\langle m,r \rangle$ only if $\langle m,r \rangle \in A$; so all the permanent requirements containing $\langle m,r \rangle$ are active at step s. Thus by (C), $\langle m,r \rangle \in A$ iff no n-requirement x with $n \leq m$ which is active at step s and contains $\langle m,r \rangle$ is permanent. In view of (A) and the induction hypothesis, we can test whether or not this is the case.

Now we consider what can be proved if we do not assume that C is piecewise recursive. We can prove just as above that A is recursive in C. In proving (D), piecewise recursiveness was only used to show that $C^{(m)}$ is recursive for $m < n$. Hence if we assume that $C^{(m)}$ is recursive for $m < n$, we can conclude that $C^{(n)} - A^{(n)}$ is finite.

Now suppose that $C^{(m)}$ is recursive for every $m > 0$. We can then still prove that A is a thick subset of C and that $D \not\leq_R A$, provided that we assume that D is strongly simple and $D \not\leq_R C^{(0)}$. The only change required in the proof is in (D). Since $C^{(0)}$ may not be recursive, we can only conclude that $k \in E_n^s$ is recursive in $C^{(0)}$ and hence that $E_n$ is RE in $C^{(0)}$. Since D is strongly simple and $D \not\leq_R C^{(0)}$, this is enough to insure that $E_n$ is not an infinite subset of $D^c$; and this is all that is needed. Again, if we assume that $C^{(m)}$ is recursive for $0 < m < n$, we can conclude that $C^{(n)} - A^{(n)}$ is finite.

Now let us consider the dependence of A upon C. In order to carry out the construction, we need an index J of C and K of D.  An index I of A can then be described as follows: computing according to I with the input $\langle m,r \rangle$ consists of carrying out the construction until $\langle m,r \rangle$ is put into A and then giving the output 0.  If we assume the index K of D is fixed, then the index I of A is a recursive function of the index J of C.

We summarize these results as follows.  Let D be a simple set in N.  Then there is a recursive function F from $\mathrm{Alg}(N \times N, N)$ to $\mathrm{Alg}(N \times N, N)$ such that if $C = W_J$ and $A = W_{F(J)}$, then: (a) A is recursive in C; (b) if $C^{(m)}$ is recursive for $m < n$, then $C^{(n)} - A^{(n)}$ is finite; (c) if C is piecewise recursive, then $D \not\leq_R A$.  If D is strongly simple and $D \not\leq_R C^{(0)}$, then: (b') if $C^{(m)}$ is recursive for $0 < m < n$, then $C^{(n)} - A^{(n)}$ is finite; (c') if $C^{(m)}$ is recursive for $m > 0$, then $D \not\leq_R A$.

## 17. Index Sets

We shall now use the results of the last section to evaluate the degrees of certain sets.

The <u>index set</u> of a, designated by $Ix(a)$, is the set of $I \in Alg(N, N)$ such that $dg\ W_I = a$. Of course $Ix(a) = \emptyset$ if a is not RE. We shall show that $dg(Ix(a)) = a^3$ if a is RE.

If $dg\ G = dg\ H$, then the same sets are recursive in G as are recursive in H; so the $\Sigma_n[G]$ ($\Pi_n[G]$) sets are the same as the $\Sigma_n[H]$ ($\Pi_n[H]$) sets. This justifies the following definition: a set is $\Sigma_n[a]$ ($\Pi_n[a]$) if it is $\Sigma_n[H]$ ($\Pi_n[H]$) where H is a function of degree a. The desired result on degrees of index sets is then implied by the following theorem.

<u>Index Set Theorem</u> (Yates). If a is RE, then $Ix(a)$ is a complete $\Sigma_3[a]$ set.

In §3, we showed that if F is recursive in G and G is recursive in H, then F is recursive in H. The proof showed how to obtain an algorithm for F in H from an algorithm for F in G and an algorithm for G in H. Thus there is a recursive function L such that if $[J]^H$ and $[I]^{[J]^H}$ are total, then $[I]^{[J]^H} = [L(I,J)]^H$.

Now we show that if a is RE, then $Ix(a)$ is $\Sigma_3[a]$. Let A be an RE set in N of degree a. With L as above:

$$I \in Ix(a) \leftrightarrow dg\ W_I = dg\ A$$
$$\leftrightarrow \exists J \exists K(W_I = [J]^A\ \&\ A = [K]^{W_I})$$
$$\leftrightarrow \exists J \exists K(W_I = [J]^A\ \&\ A = [L(K,J)]^A).$$

Thus we need only show that $W_I = [J]^A$ and $A = [L(K,J)]^A$ are $\Pi_2[A]$. Now

$$W_I = [J]^A \leftrightarrow \forall n((W_I(n) \ \& \ [J]^A(n) = Tr)$$
$$(\neg W_I(n) \ \& \ [J]^A(n) = Fa)).$$

Since $W_I(n)$ is RE and $[J]^A(n) = k$ is RE in A, both are $\Sigma_1[A]$. It follows that each of $W_I(n)$, $[J]^A(n) = Tr$, $\neg W_I(n)$, and $[J]^A(n) = Fa$ is $\Pi_2[a]$; so $W_I = [J]^A$ is $\Pi_2[A]$. We treat $A = [L(K,J)]^A$ similarly.

<u>Lemma 1</u>. If A is RE and B is $\Pi_1[a]$, then $x \in B \leftrightarrow \forall y(\langle x,y \rangle \in C)$ where C is recursive in A and RE.

<u>Proof</u>. We have $x \in B \leftrightarrow \forall z(\langle x,z \rangle \in D)$ where D is recursive in A. By the Modulus Lemma, $D = \lim D_n$ where $\{D_n\}$ is recursive and has a modulus H recursive in A. Then

$$x \in B \leftrightarrow \forall z(\lim D_n(x,z) = Tr)$$
$$\leftrightarrow \forall z \forall s \exists n(n > s \ \& \ \langle x,z \rangle \in D_n)$$

(since $\lim D_n(x,z)$ always exists). Let

$$\langle x,z,s \rangle \in C' \leftrightarrow \exists n(n > s \ \& \ \langle x,z \rangle \in D_n).$$

Then C' is RE. In the definition of C', we may replace $\exists n$ by $\exists n \leq H(x,z)$. This shows that C' is recursive in H and hence in A. Now let $Y = Z \times N$, $F(z,n) = z$, $G(z,n) = n$, and

$$\langle x,y \rangle \in C \leftrightarrow \langle x,F(y),G(y) \rangle \in C'.$$

Then C has all the desired properties. Q.E.D.

<u>Lemma 2</u>. Let A be RE and let B be $\Sigma_2[A]$. Then there is a C, recursive in A and RE, such that $C^{(x)}$ is recursive if $x \in B$

and A is recursive in $C^{(x)}$ if $x \notin B$.

  Proof. We suppose that A is a set in N. By Lemma 1
$$x \in B \leftrightarrow \exists t \, \forall s (\langle x,t,s \rangle \in D)$$
where D is recursive in A and RE. Let
$$\langle x,m,n \rangle \in C \leftrightarrow (\exists t \leq m)(\forall s \leq n)(\langle x,t,s \rangle \in D) \vee m \in A.$$
Then C is recursive in A and RE.

  Fixing x, let $E = C^{(x)}$. If $\langle m,n \rangle \in E$, then $\langle m,n' \rangle \in E$ for all $n' < n$. Hence for each m, $E^{(m)}$ is finite or equal to N. If $x \in B$, there is a t such that $\forall s (\langle x,t,s \rangle \in D)$. If $m \geq t$, then $E^{(m)} = N$. Hence in this case E is recursive. If $x \notin B$, then for each t there is an s such that $\langle x,t,s \rangle \notin D$. Hence, given m, $(\exists t \leq m)(\forall s \leq n)(\langle x,t,s \rangle \in D)$ is false for large n. Thus $E^{(m)} = N \leftrightarrow m \in A$; so $m \in A^c \leftrightarrow \exists r (\langle m,r \rangle \notin E)$. This shows that $A^c$ is RE in E. But A is RE, hence RE in E; so A is recursive in E by the Complementation Theorem. Q.E.D.

  We can now complete the proof of the Index Set Theorem. Let a be RE; we must show that every $\Sigma_3[a]$ set is reducible to Ix(a). First we let $a = 0$. (This case is due to Rogers.) Let A be an RE set in N of degree $0'$. If B is $\Sigma_3$, then B is $\Sigma_2[A']$ for some A' which is $\Sigma_1$ and hence recursive in A; so B is $\Sigma_2[A]$. Choose C as in Lemma 2. Since A is not recursive, $x \in B$ iff $C^{(x)}$ is recursive. By the Parameter Theorem, there is a recursive F such that $C^{(x)} = W_{F(x)}$. Then $x \in B \leftrightarrow F(x) \in Ix(0)$, proving that B is reducible to Ix(0).

Now let $a > 0$. By Theorem 1 of §12, there is a simple set D of degree $a$. Let B be $\Sigma_3[D]$. Then
$$x \in B^c \leftrightarrow \forall k(\langle x,k \rangle \in C)$$
where C is $\Sigma_2[D]$. By Lemma 2, there is an E, recursive in D and RE, such that

(1)     $\langle x,k \rangle \in C \rightarrow E^{(x,k)}$ is recursive,

(2)     $\langle x,k \rangle \notin C \rightarrow D$ is recursive in $E^{(x,k)}$.

Choose a recursive G by the Parameter Theorem so that $E^{(x)} = W_{G(x)}$; and let F be as at the end of §16. Set $H(x) = F(G(x))$. We complete the proof by showing that $x \in B \leftrightarrow H(x) \in Ix(a)$.

Setting $A_x = W_{H(x)}$, we have: (a) $A_x$ is recursive in $E^{(x)}$; (b) if $E^{(x,m)}$ is recursive for $m < k$, then $E^{(x,k)} - A_x^{(k)}$ is finite; (c) if $E^{(x)}$ is piecewise recursive, then $D \not\leq_R A_x$. Since $E \leq_R D$, (a) implies that $dg(A_x) \leq a$.

Let $x \in B$. Choose k minimal such that $\langle x,k \rangle \notin C$. For $m < k$, $E^{(x,m)}$ is recursive by (1); so $E^{(x,k)} - A_x^{(k)}$ is finite by (b). From this and (2), D is recursive in $A_x^{(k)}$; so $dg\, A_x = a$ and $H(x) \in Ix(a)$. Now let $x \notin B$. By (1), $E^{(x)}$ is piecewise recursive; so by (c), $D \not\leq_R A_x$. Thus $dg\, A_x < a$ and $H(x) \notin Ix(a)$. Q.E.D.

We have actually proved a little bit extra. Let $Ix(b,a)$ be the set of $I \in Alg(N,N)$ such that $b \leq dg\, W_I < a$. We have shown that if $a$ is RE and $a \neq 0$, then every $\Sigma_3[a]$ set is reducible to $Ix(a)$, $Ix(0,a)$.

We shall improve this result to the following: if a and b are RE and $b < a$, then every $\Sigma_3[a]$ set is reducible to $Ix(a)$, $Ix(b,a)$. By the remark of §12, there is a strongly simple set D of degree a. Let B be $\Sigma_3[D]$, and take C and E as above. Let D' be an RE set in N of degree b, and define

$$\langle x,k,n \rangle \in E' \leftrightarrow (k = 0 \ \& \ n \in D') \lor (k > 0 \ \& \ \langle x,k-1,n \rangle \in E).$$

Then E' is RE and is recursive in D',E. Since $b < a$, E' is recursive in D. Choose a recursive G such that $W_{G(x)} = E'^{(x)}$, and set $H(x) = F(G(x))$. We show that $x \in B \to H(x) \in Ix(a)$ and $x \notin B \to H(x) \in Ix(b,a)$.

We have $E'^{(x,0)} = D'$ and $E'^{(x,k+1)} = E^{(x,k)}$. Since $b < a$, $D \not\leq_R E'^{(x,0)}$. Hence, setting $A_x = W_{H(x)}$, we have: (a) $A_x$ is recursive in $E'^{(x)}$; (b) if $E^{(x,m)}$ is recursive for $m < k$, then $E^{(x,k)} - A_x^{(k+1)}$ is finite; (c) if $E^{(x)}$ is piecewise recursive, then $D \leq_R A_x$. Just as above, we can combine these results with (1) and (2) to show that dg $A_x = a$ if $x \in B$ and dg $A_x < a$ if $x \notin B$. Moreover, $E'^{(x,0)} - A_x^{(0)}$ is finite; so $D' = E'^{(x,0)} \leq_R A_x^{(0)} \leq_R A_x$. Thus $b \leq$ dg $A_x$. Hence our result is proved.

We shall use this result to prove another theorem about RE degrees. First we prove an important result which shows that in defining an RE set, we may use an index of the set we are defining.

<u>Recursion Theorem</u> (Kleene). If A is an RE set in Alg(X,N) ×

X, then there is an I such that $x \in W_I \leftrightarrow \langle I,x \rangle \in A$ for all x.

*Proof.* We use the Isomorphism Theorem to identify $\text{Alg}(X,N)$ and $\text{Alg}(\text{Alg}(X,N) \times X, N)$. Then $\langle J,x \rangle \in W_J$ is an RE relation of J,x; so by the Parameter Theorem, there is a recursive function F such that $x \in W_{F(J)} \leftrightarrow \langle J,x \rangle \in W_J$ for all J and x. Choose J so that $\langle I,x \rangle \in W_J \leftrightarrow \langle F(I),x \rangle \in A$ for all I and x; and set $I = F(J)$. Then

$$x \in W_I \leftrightarrow \langle J,x \rangle \in W_J$$
$$\leftrightarrow \langle F(J),x \rangle \in A$$
$$\leftrightarrow \langle I,x \rangle \in A. \quad \text{Q.E.D.}$$

*Corollary* (*Fixed Point Theorem*). If F is a recursive function from $\text{Alg}(X,N)$ to $\text{Alg}(X,N)$, then there is an I such that $W_I = W_{F(I)}$.

*Proof.* Define the A of the Recursion Theorem by

$$\langle I,x \rangle \in A \leftrightarrow x \in W_{F(I)}. \quad \text{Q.E.D.}$$

*Density Theorem* (Sacks). If a and b are RE degrees such that $b < a$, then there is an RE degree c such that $b < c < a$.

*Proof.* Since $\text{Ix}(b)$ is $\Sigma_3[b]$ and hence $\Sigma_3[a]$, there is a recursive F such that $I \in \text{Ix}(b) \to F(I) \in \text{Ix}(a)$ and $I \notin \text{Ix}(b) \to F(I) \in \text{Ix}(b,a)$. Choose I as in the Fixed Point Theorem. If $I \in \text{Ix}(b)$, then $b = \text{dg } W_I = \text{dg } W_{F(I)} = a$, a contradiction. Thus $I \notin \text{Ix}(b)$; so $\text{dg } W_I \neq b$ and $b \leq \text{dg } W_{F(I)} < a$. Thus we may take $c = \text{dg } W_I = \text{dg } W_{F(I)}$. Q.E.D.

## 18. Branching Degrees

Recent work in degree theory has been largely concerned with the RE degrees. We consider here only one of the problems which has been studied.

An RE degree a is <u>branching</u> if there are RE degrees b and c different from a such that a is the glb of $\{b,c\}$. Clearly 0' is not branching. The existence of branching degrees is given by our first result.

<u>Theorem 1</u> (Yates). The degree 0 is branching.

<u>Proof</u>. We shall construct two non-recursive RE sets A and B in N such that $\{dg\ A,\ dg\ B\}$ has glb 0. The conditions to be satisfied are:

($1_I$) $A \neq [I]$;

($2_I$) $B \neq [I]$;

($3_{I,J}$) If $[I]^A = [J]^B = F$, then F is recursive.

Let $\{R_n\}$ be a listing of the space of conditions.

Our idea for insuring ($3_{I,J}$) is to insure that if $F = [I]^A = [J]^B$ and if $[I]^{A^s}(k) = [J]^{B^s}(k) = 1$ for some s, then $F(k) = 1$. This will clearly be the case if for every $t \geq s$, either $[I]^{A^t}(k) = 1$ or $[J]^{B^t}(k) = 1$. Hence when we discover that $[I]^{A^s}(k) = 1$, we create a requirement; but we do not use it to keep elements out of A at a step at which there is an active requirement corresponding to $[J]^{B^t}(k) = 1$. An exception to this is provided by certain elements of the requirement, called <u>key</u> elements, which cannot be put into A while the requirement is active. The purpose of this is to insure that we do not create infinitely many requirements for a single argument k.

Since we may have infinitely many n-requirements, we are again faced with the problem of insuring that a number is not kept out of A (or B) by infinitely many temporary requirements. We solve this problem by a different method from that used in §16. At each step s, certain numbers are put into A (or B) despite the existence of an n-requirement which would normally keep them out of A. We describe these numbers exactly later; roughly, they are numbers which at many previous steps have been kept out of A by m-requirements with $m \leq n$.

We now turn to some definitions. If $R_n$ is $(3_{I,J})$, we may create n-requirements. Each such requirement is either a requirement <u>for</u> A or a requirement <u>for</u> B. Below, we often consider only requirements for A; it is understood that everything also holds when A and B are interchanged.

Each requirement will have an <u>argument</u> and a <u>value</u>. If at step s we create an n-requirement x for A with argument k and value i, we will have $x \cap A^s = \emptyset$. For $t > s$, x is <u>active</u> at step t if $x \cap A^t = \emptyset$; otherwise, x is <u>inactive</u> at step t. We say x is <u>effective</u> at step t if x is active at step t and no n-requirement for B with argument k and value i is active at step t; otherwise, we say x is <u>ineffective</u> at step t. An element r of x is a <u>key</u> element if r is in row m and $m > k + n$.

We now define finite sets $P_s^A(n,k)$ and $Q_s^A(n)$ by induction on s. The idea is that $P_s^A(n,k)$ is the set of numbers which we do not allow an n-requirement for A with argument k to keep out of A at step s; and $Q_s^A(n)$ is the set of numbers kept out of A at step s by m-requirements for A with $m \leq n$. Precisely, $r \in P_s^A(n,k)$ if r belongs to an n-requirement for A active at step s and $r \in Q_t^A(n)$ for all t such that $k \leq t < s$ and t is in the same row as r; and $r \in Q_s^A(n)$ if r belongs to an m-requirement for A with argument k which is effective at step s, $m \leq n$, and $r \notin P_s^A(m,k)$.

We now describe step s. Let s be in row n, and first suppose that $R_n$ is $(1_I)$. We look for a number $r \leq s$ such that: (a) no number in $A^s$ is in row n; (b) r is in row n; (c) $[I]_s(r) = Fa$; (d) r is not a key element in an active requirement for A; (e) $r \notin Q_s^A(n)$. If there is such an r, we put the smallest such r into A.

If $R_n$ is $(2_I)$, we merely interchange A with B in the above. Now suppose that $R_n$ is $(3_{I,J})$. Then step s has two parts, an A-part and a B-part. We describe the A-part; the B-part is similar, except that we interchange A with B and I with J.

Let k be the smallest number not the argument of an active n-requirement for A. We do nothing unless $k \in W_{I,s}^{A^s}$ and no effective n-requirement has an argument < k. In this case, we create an n-requirement for A with argument k and value $[I]^{A^s}(k)$, consisting of all numbers used negatively in the computation of $[I]^{A^s}(k)$.

We prove a sequence of results leading to the conclusion that all of the conditions are satisfied.

We define _permanent_ and _temporary_ requirements as usual. A requirement is _essential_ if it is effective at step s for all sufficiently large s; otherwise, it is _inessential_.

(A) For each n and k, there are only finitely many n-requirements for A with argument k.

If a number r in an active n-requirement for A with argument k is put into A, then r is not a key element in the requirement, so r is in row m where $m \leq n + k$. For each m, at most one number in row m is put into A. These facts imply that there are only finitely many temporary n-requirements for A with argument k; and there is at most one permanent n-requirement for A with argument k.

(B) If x is an inessential requirement, then for all sufficiently large s, x is ineffective at step s.

Let x be an n-requirement for A with argument k. If the result is false, then there are infinitely many s such that x is effective at step s and ineffective but active at step s+1. This implies that an n-requirement for B with argument k is created at step s; and this cannot happen infinitely often by (A).

(C) For each n, there are at most finitely many essential n-requirements for A.

Suppose that there is an essential n-requirement for A with argument k. Then after some stage no n-requirement for A with an argument $> k$ is created. Hence by (A), there are only finitely many n-requirements for A.

We let $r \in Q^A(n)$ if $r \in Q^A_s(n)$ for all sufficiently large s in the same row as r.

(D) If n is the smallest number such that $r \in Q^A(n)$, then r belongs to an essential n-requirement.

Suppose not. Choose $s_0$ so that $r \in Q^A_s(n)$ if $s \geq s_0$ and s is in the same row as r. By (A), r belongs to only finitely many n-requirements for A with arguments $< s_0$. By hypothesis, they are all inessential. Hence by (D), there is an $s_1 \geq s_0$ such that for $s \geq s_1$, all of these requirements are ineffective.

Since $r \in Q^A(n)$ and $r \notin Q^A(n-1)$, there are infinitely many s in the same row as r such that $r \in Q^A_s(n) - Q^A_s(n-1)$. Pick such an s with $s \geq s_1$. At step s, r belongs to an effective n-requirement x for A with argument k and $r \notin P^A_s(n,k)$. Hence there is a t in the same row as r such that $k \leq t < s$ and $r \notin Q^A_t(n)$. By the choice of $s_0$, $k \leq t < s_0$. But this contradicts the choice of $s_1$.

(E) $Q^A(n)$ is finite.

This follows from (D) and (C).

We now show that $(1_\perp)$ holds. Let $(1_I)$ be $R_n$. If a number r in row n is put into A, then $[I](r) = Fa$; so $(1_I)$ holds. Now suppose that A contains no number in row n. We assume that $A = [I]$ and derive a contradiction. Choose r in row n so that r is not a key element in a requirement for A and $r \notin Q^A(n)$; this is possible by (A) and (E). Since $r \notin A$, $[I](r) = Fa$. Hence we can choose s in row n so large that $s \geq r$, $[I]_s(r) = Fa$, and $r \notin Q^A_s(n)$. Then at step s a number in row n is put into A, a contradiction.

(F) For each n and k, $P^A_s(n,k) \subset Q^A(n)$ for all sufficiently large s.

Since $P^A_s(n,k)$ is finite and decreases with increasing s, it suffices to show that if $r \in P^A_s(n,k)$ for all s, then $r \in Q^A(n)$. This is immediate.

Now we turn to $(3_{I,J})$. Let $(3_{I,J})$ be $R_n$, and assume that $[I]^A = [J]^B = F$. Then every permanent n-requirement with argument k has value $F(k)$. We show that for each k, there are permanent inessential n-requirements for A and for B with argument k. We use induction on k. By the induction hypothesis, (B), and (A), we have for all sufficiently large s: (a) for each $j < k$, the permanent n-requirements for A and B with argument j are ineffective at step s; (b) every temporary n-requirement for A or B with argument k is inactive at step s; (c) $k \in W^{A^s}_{I,s} \cap W^{B^s}_{J,s}$. Choose such an s in row n. Then at step s, either there is an active n-requirement for A with argument k or such a requirement is created. In either case, the requirement is permanent by (b). Similarly, there is a permanent n-requirement for B with argument k. Since both of these requirements have value $F(k)$, they are inessential.

Choose $s_0$ so that $Q^A(n) \cap A \subset A^{s_0}$; every member of A which is in a row $m \leq n$ is in $A^{s_0}$; and the corresponding results with A replaced by B also hold. Using $Q^A(n)$ and $s_0$, we shall compute $F(k)$ from k.

Let k be given. Choose $s_1 \geq s_0$ so that $P^A_{s_1}(n,k) \subset Q^A(n)$ and there is an n-requirement x for A with argument k which is active and ineffective at step $s_1$. Such an $s_1$ exists by (F) and the above, and it can be found by trial. Let i be the value of x; we show that $F(k) = i$.

We know that for sufficiently large x, there will be at step s active n-requirements for A and B, both with argument k and value $F(k)$. Hence it suffices to show that for each step $s \geq s_1$, there is an active n-requirement for A or B with argument k and value i. This holds for $s = s_1$; so it suffices to show that if it holds for $s \geq s_1$, then it holds for s+1.

Suppose that at step s there are active n-requirements for both A and B with argument k and value i. Since numbers cannot be put in both A and B at step s, one of these requirements is active at step s+1.

Now suppose, say, that at step s there is an active n-requirement for A with argument k and value i, but not for B. This requirement y is then effective at step s; and it will suffice to derive a contradiction from the assumption that a number $r \in y$ is put into A at step s. Let r be in row m. Then $r \notin Q_s^A(m)$; and from the choice of $s_0$, $n < m$. Since r is in the effective requirement y, $r \in P_s^A(n,k)$; so $r \in Q^A(n)$ by the choice of $s_1$. But this contradicts the choice of $s_0$. Q.E.D.

We now prove that there are non-branching RE degrees other than $0'$.

<u>Theorem 2</u> (Lachlan). If d is a non-zero RE degree, then there is a non-branching RE degree a such that $a \leq d$.

<u>Lemma</u>. Let E be an infinite set in N which is RE in A. Let G be a function from N to N which is recursive in A. Let $I \in Alg(N,N)$ be such that $dg\, A < dg\, W_I$. Let E' be the set of $r \in E$ such that some number $\leq r$ is in $W_I - W_I^{G(r)}$. Then E' is RE in A and infinite.

<u>Proof</u>. Since
$$r \in E' \leftrightarrow r \in E\ \&\ (\exists s \leq r)(s \in W_I\ \&\ s \notin W_I^{G(r)}),$$
E' is RE in A. Suppose that E' is finite, so that E - E' is RE in A and infinite. By the Selection Theorem there is a function F, recursive in A, such that $n \leq F(n)$ and $F(n) \in E - E'$ for all n. Then we have
$$n \in W_I \leftrightarrow n \in W_I^{G(F(n))};$$

so $W_I \leq_R A$. This contradicts $\operatorname{dg} A < \operatorname{dg} W_I$. Q.E.D.

Now we turn to Theorem 2. Let D be an RE set in N of degree d. Since D is non-recursive, it is infinite; so D has a listing F. We construct an RE set A in N and set $a = \operatorname{dg} A$. We make A recursive in D, so that $a \leq d$.

To make A non-branching, we must insure that whenever

(1)    $\operatorname{dg} A < \operatorname{dg} W_I$ & $\operatorname{dg} A < \operatorname{dg} W_J$,

then $\operatorname{dg} A$ is not the glb of $\{\operatorname{dg} W_I, \operatorname{dg} W_J\}$; i.e., there is a $B_{I,J}$ such that $B_{I,J} \leq_R W_I$, $B_{I,J} \leq_R W_J$, and $B_{I,J} \not\leq_R A$. Our chief effort in the construction is devoted to making $B_{I,J} \not\leq_R A$. Thus we have the conditions:

$$(1_{I,J,K}) \quad B_{I,J} \neq [K]^A.$$

Let $\{R_n\}$ be a listing of the space of conditions.

We say that row n is an (I,J)-row if $R_n$ is $(1_{I,J,K})$ for some K. If k is in row n, we write $\operatorname{rw}(k)$ for n. We let $E_{I,J}$ be the set of $\operatorname{rw}(k)$ for $k \in A$ and $\operatorname{rw}(k)$ in an (I,J)-row. Thus if $R_n$ is $(1_{I,J,K})$, we try to insure $R_n$ by finding a k such that $\operatorname{rw}(k)$ is in row n and $[K]^{A^s}(\operatorname{rw}(k)) = Fa$, and then putting k into A and insuring that $[K]^A(\operatorname{rw}(k)) = Fa$. For the latter purpose, we may introduce an n-requirement with __index__ k.

If an n-requirement x with index k is created at step s, we will have $x \cap A^s = \emptyset$. For $t > s$, x is __active__ at step t if $x \cap A^t = \emptyset$; otherwise, x is __inactive__ at step t. We say x is __effective__ at step t if it is active at step t and $k \in A^t$.

Now we describe step s. First suppose that $s = 2t$. For each n, we put into A the smallest k satisfying the following conditions, if there is such a k: (a) k is the index of an active n-requirement; (b) there is no effective n-requirement; (c) k is in no active m-requirement with $m < n$; (d) $k \geq F(t)$.

Now suppose that $s = 2t+1$, and let t be in row n. If there is an effective n-requirement, we do nothing. Otherwise, we look for a number $k \leq s$ such that:

(i) k is not the index of a previously created n-requirement;

(ii) $rw(k)$ is in row n;

(iii) $[K]_s^{A^s}(rw(k)) = Fa$;

(iv) k is not used in the computation of $[K]^{A^s}(rw(k))$;

(v) some number $\leq rw(k)$ is in $W_I^s - W_I^k$;

(vi) some number $\leq rw(k)$ is in $W_J^s - W_J^k$.

If such a k exists, we pick the smallest such k, and create an n-requirement with index k consisting of all numbers used negatively in the computation of $[K]^{A^s}(rw(k))$.

We first show that $A \leq_R D$. Let an oracle for D and a number k be given. Choose t so that every number $\leq k$ in D is among $F(0), F(1), \ldots, F(t-1)$. Then k cannot be put into A at step 2t or later; so $k \in A \leftrightarrow k \in A^{2t}$.

We note that by (i) and (ii), no number is the index of more than one requirement. If an n-requirement y is active at step s and a number r in y is put into A at step s, then r is the index of an active m-requirement x; and $m \leq n$ by (c). We say that x <u>kills</u> y at step s. By (iv), $x \neq y$.

We show that if an n-requirement x kills an n-requirement y at step s, then the index k of y is not in A. By (b), y is not effective at step s; so $k \notin A^s$. Since x is the only n-requirement whose index is put into A at step s, k is not put into A at step s. After step s, the only requirement with index k is inactive; so k is never put into A.

We define <u>permanent</u> and <u>temporary</u> requirements as usual. A requirement is <u>essential</u> if it is permanent and its index belongs to A; this means that it is effective at step s for all sufficiently large s.

We now show by induction on n that there are only finitely many n-requirements. If there is an essential n-requirement, then after some step no more n-requirements are created; so we suppose that there are no essential n-requirements. By the induction hypothesis, only finitely many n-requirements are killed by m-requirements with $m < n$. An n-requirement which is killed by an n-requirement cannot have its index in A, as we saw above. We conclude that only finitely many n-requirements have their index in A. But if x is killed by y, then the index of y is in A. It follows that there are only finitely many temporary n-requirements.

Suppose that there are infinitely many n-requirements. We obtain the desired contradiction by proving that D is recursive. Choose $s_0$ so that every temporary n-requirement is inactive at step $s_0$ and every n-requirement whose index is in an m-requirement with $m < n$ is created before step $s_0$. Let r be given. There is an $s \geq s_0$ such that an n-requirement with an index $k \geq r$ is created at step s; and we can find s. If $t \geq s$, then no n-requirement is effective at step 2t; x is active at step 2t; k is in no m-requirement with $m < n$; and no index of an n-requirement is put into A at step 2t. We conclude that $F(t) > k \geq r$ for $t \geq s$; so $r \in D$ iff r is among $F(0), F(1), \ldots, F(s-1)$.

We now assume (1). First we show that $B_{I,J} \leq_R W_I$. We have $r \in B_{I,J}$ iff r is in an $(I,J)$-row and some k in row r is in A. Since $A \leq_R W_I$, it will suffice to show that we can bound this k with the aid of an oracle for $W_I$. By (v), some number $\leq r$ must be in $W_I - W_I^k$. Thus if we choose s so large that every number $\leq r$ in $W_I$ is in $W_I^s$, then $k \leq s$. A similar proof shows that $B_{I,J} \leq_R W_J$.

It remains to show that $B_{I,J} \not\leq_R A$. We assume that $B_{I,J} = [K]^A$ and derive a contradiction. Choose n so that $R_n$ is $(1_{I,J,K})$. Suppose first that there is an essential n-requirement. If k is its index, then $[K]^A(rw(k)) = Fa$ and $rw(k) \in B_{I,J}$. This contradicts our assumption.

Now suppose that there is no essential n-requirement. Let $k_0$ be larger than any index of an n-requirement; and let $r_0$ be larger than $rw(k)$ for all $k < k_0$. Let E be the set of numbers $\geq r_0$ in row n. Then E is recursive and infinite, and $E \cap B_{I,J} = \emptyset$. Thus $r \in E \rightarrow [K]^A(r) = Fa$.

For $r \in E$, let $G(r)$ be the smallest number $k \geq k_0$ in row r greater than every number used in the computation of $[K]^A(r)$; for $r \notin E$, let $G(r) = 0$. Then G is recursive in A; and if $r \in E$, then $G(r) = k$ satisfies (i)-(iv) for all sufficiently large s. By two uses of the lemma, there is an $r \in E$ such that both $W_I - W_I^{G(r)}$ and $W_J - W_J^{G(r)}$ contain a number $\leq r$. Then for all sufficiently large s, $G(r)$ satisfies (i)-(vi). Since there is no essential n-requirement, there is no effective n-requirement at step s for all sufficiently large s. This implies that infinitely many n-requirements are created, which is a contradiction. Q.E.D.

Lachlan has also shown that there is a branching degree other than 0; the proof is similar to the proof of Theorem 1. Thus there does not seem to be any very simple characterization of branching degrees.